WAYS

AND

HIGHWAYS.

THE STATUTES OF NEW YORK

IN RELATION TO

HIGHWAYS, BRIDGES, FERRIES, PLANK ROADS AND TURNPIKE ROADS,

WITH COMMENTARIES:

ALSO

AN APPENDIX,

CONTAINING

ALL THE FORMS AND PRECEDENTS

TO BE USED UNDER SAID STATUTES.

BY WILLIAM S. BISHOP,
COUNSELLOR AT LAW.

ROCHESTER, N. Y.:
PRINTED BY LEE, MANN & CO., AMERICAN OFFICE.
1851.

ADVERTISEMENT.

The former editions of this work having been exhausted, the Editor has prepared another with much care and attention, and he hopes it may be found convenient and useful to the public. The last edition has been carefully revised and corrected, while at the same time many new and important additions have been made.

To the thousands of public officers who have the control and management of the Highways in this State, it is thought that this manual, in its present improved and enlarged form, may be peculiarly serviceable.

Whenever the Highway Law has received any Judicial construction by the Courts, those cases have been, with no inconsiderable labor, collected under the sections to which they relate ; so that the decisions of our Courts may be seen at a glance.

It is hoped, also, that this work may be found very convenient to the profession ; as the Editor has not omitted, as he believes, any decisions bearing upon this subject ; and they can be easily referred to without the trouble of turning over our Reports and Digests, by merely looking at the appropriate sections.

The "Forms" are prepared with the utmost care and attention.

The chapter on "Ways," it is believed contains a brief reference to all the American Law, on that subject applicable to this State, and also a reference to several English cases, which throw light on that branch of law. As there is not, within our knowledge, an American Treatise on "Ways" this portion of the work, we are inclined to believe, will be of no slight value to the profession.

There is also a collection of the cases, of the "Law of the Road."

The laws in relation to Plank Roads and Turnpikes, with amendments which have passed the Legislature, are added, with forms.

The Highway Act has long been the Law of this State, and in its original form is well known to most of the public officers having charge of that important subject. Under it thousands and thousands of miles of Roads have been constructed by the peaceful citizen, and the burden slightly felt. Changes and innovations affecting such great interests and so many individuals, should be adopted with unusual caution. Injudicious and hasty legislation would multiply litigation, and be productive of much evil, by creating disturbances and ill-will, in districts and neighborhoods.

OF

HIGHWAYS, BRIDGES & FERRIES.

TITLE 1.—OF HIGHWAYS AND BRIDGES.
TITLE 2.—OF THE REGULATIONS OF FERRIES.

TITLE I.

OF HIGHWAYS AND BRIDGES.

ART. 1.—Of the officers entrusted with the care and superintendence of highways and bridges; and their general powers and duties.
ART. 2.—Of the persons liable to work on highway, and the making of assessment therefor.
ART. 3.—Of the duties of overseers in regard to the performance of labor upon highways, and of the performance of such labor, or the commutation therefor.
ART. 4.—Of the laying out of public and private roads, and of the alteration or discontinuance thereof.
ART. 5.—Regulations and penalties concerning the obstructing of highways, and encroachments thereon.
ART. 6.—Of the erection, repairing, and preservation of bridges.
ART. 7.—Of miscellaneous provisions of a general nature.

ARTICLE FIRST.

Of the Officers entrusted with the Care and Superintendence of Highways and Bridges; and their general Powers and Duties.

SEC. 1. Commissioners of highways to have care of highways and bridges; their duties.
2. They have power to lay and discontinue roads.
3. To account to board of auditors of town accounts.
4. To deliver statement of improvements necessary of roads and bridges.
5. They shall cause mile-stones to be erected.
6 & 7. Duties of overseers of highways.
8. When to make a new assessment.
9. Commissioners of highways to cause guide-posts to be erected.
10. Overseers of highways to keep them in repair.
11. Commissioners may procure, with fines and commutation moneys, a scraper, &c.

Commissioners, their duty

§ 1. The commissioners of highways in the several towns in this state, shall have the care and superintendence of the highways and bridges therein: and it shall be their duty,

1. To give directions for the repairing of the roads and bridges, within their respective towns;

2. To regulate the roads already laid out, and to alter such of them as they, or a majority of them, shall deem inconvenient.

3. To cause such of the roads used as highways, as shall have been laid out, but not sufficiently described, and such as shall have been used for twenty years, but not recorded, to be ascertained, described and entered of record in the town clerk's office:

4. To cause the highways, and the bridges which are, or may be erected over streams intersecting highways, to be kept in repair:

5. To divide their respective towns into so many road districts as they shall judge convenient, by writing under their hands, to be lodged with the town clerk, and by him to be entered in the town book; such division to be made annually, if they shall think it necessary, and in all cases to be made at least ten days before the annual town meeting:

6. To assign to each of the said road districts, such of the inhabitants liable to work on highways, as

they shall think proper, having regard to proximity of residence as much as may be; and

7. To require the overseers of highways, from time to time, and as often as they shall deem necessary, to warn all persons assessed to work on highways, to come and work thereon, with such implements, carriages, cattle or sleds, as the said commissioners, or any of them may direct.

By the act of May 10, 1845, *Session Laws, page* 183 ; as amended by the act of December, *Laws of* 1847, *page* 580 ; it is provided that the several towns may choose one or three commissioners; and therefore when the word *commissioner* is used in this act, it is to be understood the same as though it read the *commissioner or commissioners of a town.* The words of the act are as follows:

"The electors of each town shall have power at their annual town meeting, to determine by resolution, whether there shall be chosen one or three highway commissioners, and the number so determined upon shall be balloted for and chosen; and if only one shall be determined upon and chosen, he shall possess all the powers, and discharge all the duties of commissoner of highways, as prescribed by law, and shall hold his office for one year. And whenever three commissioners shall be chosen in any town, they shall be divided by lot by the canvassers, upon the result of the canvass, into three classes, to be numbered one, two, three; and the term of office of the first class shall be one year, of the second, two, and of the third, three; and one commissioner only shall thereafter annually be elected in such town, who shall hold his office for three years, and until a successor shall be duly elected or appointed. But in case any commissioner shall be elected to fill a vacancy, he shall hold the office only for the unexpired term which shall become vacant."

"And if two vacancies shall be required to be filled, the canvassers shall, after the canvass, determine by lot as aforesaid, the terms they shall respectively hold. And when any vacancy shall happen by death, removal, resignation, neglect to qualify, or refusal, to serve, it shall be supplied until the next succeeding town meeting; by an appointment in writing, under the hands of any three justices of the peace, or two justices of the peace and the supervisor of the town ; and every commissioner of highways shall be authorized to administer oaths to any witnesses or juries, in proceedings which may be had by or before them; and whenever any town shall have determined upon having three commissioners, and shall desire to return two or have but one, such town shall have the power so to do by a resolution taken at an annual town meeting, and when such resolution shall have been adopted, no other commissioner shall be elected or appointed, until the term or terms of those in office, at the time of adopting such resulution, shall expire or become vacant, and they shall have power to act until their terms shall severally become vacant or expire, as fully as if the three continued in office."

"Every commissioner of highways hereafter to be elected or "appointed, shall before entering upon his duties, and within ten days "after notice of his election or appointment, execute to the supervi- "sor of his town, a bond with two sureties, to be approved by the "supervisor by an endorsement thereon, and filed with him, in the "penal sum of one thousand dollars, conditioned that he will faith- "fully discharge his duties as such commissioner, and within ten days "after the expiration of his term of office, pay over to his successor "what money may be remaining in his hands as such commissioner, "and render to such successor a true account of all moneys received "and paid out by him as such commissioner."—*Session Laws of* 1845, *page* 183.

A commissioner of highways must be an elector of the town, 1. R. S. 345, § 20.* They are to be chosen in each town at the annual town meeting. 1 R. S. 340, § 4.

The recent act makes an alteration in supplying a vacancy occasioned by *death, removal or resignation*. By the 1 R. S. 348, § 34. 35. The town clerk should call a special town meeting within eight days after the happening of the vacancy, and if the electors of the town did not within fifteen days after the happening of the vacancy fill the office, the justices of the peace of the town should make the appointment. The vacancy can now be supplied by two justices until the next annual town meeting. This provision may lead to difficulty, as two justices may appoint one person, and the other two justices in the town appoint another. The two that act first will probably be deemed to have acquired the power.

There is also a new power conferred on the commissioners, of administering oaths to witnesses and juries.—

When there is a failure to elect at the annual town meeting, in consequence of a *tie* in the votes, and the meeting adjourns without making an election, the justices of the town may make the appointment. The People vs. Van Horne, 18 Wend. 515.

Care and Superintendence.—These words confer upon the commissioners the general control and oversight of all highways and bridges in their town; but it should not be understood as giving them any power over other officers intrusted with duties in relation to highways, except as given by statute.

Highways and Bridges.—Public highways and bridges: not those made by individuals unless they have become *public* by the acts of individuals, or by the commissioners in the manner set forth in the fourth article of this act.

In this act the term *road* is used synonymously with *highway*. 9. Johns. R. 349.

SUB. 1. By this subdivision it is the duty of the commissioners and they have the authority to direct the overseer as to the general manner he shall improve the roads or bridges in his district. As for

---

* NOTE.—All reference to the Revised Statutes in this work, are to the first edition.

instance the grade of the road how drained or leveled, or in what manner a bridge shall be repaired.

The overseer is subordinate to the commissioners. The principal duties of the overseers are to warn persons to work, to see that they do work, and to collect fines and commutation money. The commissioners have the exclusive control in directing the manner in which bridges shall be repaired, and on this subject are the persons responsible to the public, and not the overseers. See notes under section 6, and *Bartlett vs. Crozier* 17, Johns. R. 452, and cases therein cited.

An overseer, however, is liable for not removing obstructions in the highway, although not specially required by the commissioners. *McFadden vs. Kingsbury*, 11 Wendell, 667.

SUB. 2. *To regulate and alter*, that is to restore the boundaries and fences of a highway to its original lines; and if it passes by or through an inconvenient place, they may change its location; but if in making the alteration, it is necessary to take improved or cultivated lands, damages must be paid, and the same course taken as on an original location of a new road.

SUB. 3. This subdivision authorizes the commissioners in their respective towns, to cause such of the roads as are not already described and recorded, to be ascertained, described, and entered of record. This authority necessarily implies and presupposes an omission of some of the requisites to the establishment of a public highway. See § 107.

In the case of *Colden vs. Thurber*, 2 *Johns. R.* 424, the commissioners of highways of Cortlandville made an order adjudging that a certain road in that town particularly designated, *had been used as a public highway for twenty years but had not been recorded*; and they thereupon proceeded *to ascertain describe and enter* the *road of record*, in the town clerk's office. Peck appealed to the judges. From the return it appeared that the commissioners in their order declared the road to be *three* rods wide, and they insisted before the judges that this was the width originally intended. It was proved before the judges that the road never was opened more than *two* rods wide. Judge BRONSON says: "This provision does not authorize the commissioners to say what was 'originally intended,' either by the owner of the soil or any one else, in relation to the width or location of the road, any further than such intention has been manifested by permitting the way to be used. It is a power in relation to the road *as it actually exists* and *has existed* for the last twenty years. It does not authorize the commissioners to create or enlarge but only to perpetuate the evidence of a public right. Both the *extent* and the *fact* of dedication depend on the user; and the public must take *secundum formam doni*. On the facts returned by the judges, it is quite clear that the commissioners exceeded their powers, and the only question is, whether Peck had a remedy by appeal?"

It does not seem to require an order or certificate to be signed by the commissioners, other than the town record. It merely directs that it shall be entered of record. It would be the better mode for the commissioners to make a written order, sign it, and have the town clerk record it. See forms, Appendix No. 1 and 2.

Sub. 4. They are not only to keep the bridges in repair, but also the causeways or entrances on to the bridges in repair, if they have sufficient money in their hands to do it.

Chancellor Kent says in 17, *Johns. R. page* 451 : "If a private "suit will lie in any case, for the recovery of damages for a broken "bridge, I should suppose the commissioners of highways are the "only persons to be sued." Their duty is to give directions, &c. This appears to be a general duty applicable to all times and in all places; yet when we come to read the details of their duty, we perceive that it does not exist absolutely, but arises only when the commissioners have money in hand from forfeitures and penalties, or which has been paid over to them under the direction of the supervisors.

It was also decided in the *People vs. Com. of Hudson,* 7 Wen. 474, that commissioners are not bound to build bridges when not in funds to defray the expense; and that a mandamus would not lie to compel them to rebuild a bridge that had been carried away.

Neither will an indictment lie against the commissioners for not repairing a bridge, unless they have funds; and an indictment against them is defective unless it avers that the defendants have funds or other means, to defray the expense of the repairs. *The People vs. Adsit,* 2 Hill's 619.

If the amount which the town is authorized to raise, is not sufficient to rebuild or repair a bridge, there must be a special act of the legislature, authorizing a sufficient amount to be raised. 7 *Wendell,* 477.

It is frequently a question of difficulty to decide what is the true legal meaning of the word "*bridge*" under this act.

While no one doubts that the structure over a large stream would be deemed a *bridge*, as the stream becomes smaller it will descend to a *culvert,* a mere *sluice-way*; and there is no strict legal line of *demarcation.* An overseer is bound no doubt to repair a *sluice-way,* while he is not bound to repair a bridge, and would be liable in damages to a person if he should fail to keep a sluice-way in repair if he had the means, and this will extend to bridges over small brooks in his district but when his liabilities and powers will cease must depend upon each particular case.

Sub. 5. See form in Appendix, No. 3.

Sub. 6. Of course, any person living directly on a road in a district will be assigned to that district; but some persons may live off from any road liable to work on highways, and they should be assigned to the district most convenient for them.

When a plank road or turnpike road is built upon the site of an old highway, it is made the duty of the commissioners of highways of the town in which such is made, to designate some district or districts within their town on which the highway labor of the inhabtants residing along the line of such plank or turnpike road, shall be performed. *See the plank road law on this subject.*

Where a road district was formed from parts of two districts, and was afterwards ordered to be discontinued, the order was held valid, though it did not expressly provide for embracing the territory to which it related within any other district; the effect was to restore the two districts to their original limits. *The People vs. Sly and others.* 4 *Hill's* 593.

SUB. 7. The duty first enjoined in this subdivision, is generally done by delivery of the tax list only. It is the duty of the commissioners, in case there is negligence on the part of the overseer, or if negligence is apprehended, to give a special requirement to the overseer.

---

§ 2. The commissioners of highways shall have power, in the manner and under the restrictions hereinafter provided, to lay out on actual survey, such new roads in their respective towns as they may deem necessary and proper; and to discontinue such old roads and highways, as shall appear to them, on the oaths of twelve freeholders of the same town, to have become unnecessary. To lay out and discontinue roads.

The commissioners have the absolute power to lay out a road in any part of their town under certain *restrictions.*

These restrictions are: 1. They cannot lay out a road without the consent of the owner, through any orchard of the growth of four years or more; nor over a garden cultivated four years; nor through any buildings; or any fixtures or erections for the purpose of trade or manufactures; or any yards or enclosures necessary for their use, nor through any enclosed, improved or cultivated lands, without the consent of the owner, or on the oath of *twelve freeholders.* See § 57, § 58, &c. Notes. No road whatever can be discontinued, without the oath of twelve freeholders.

---

§ 3. The commissioners of highways of each town, shall render to the board of town auditors, at their annual meeting for auditing the accounts of town officers, an account in writing, stating,

1. The labor assessed and performed in such town;

2. The sums received by such commissioners for fines and commutations, and all other moneys received under this chapter;

3. The improvements which have been made on the roads and bridges in their town, during the year immediately preceding such report, and an account of the state of such roads and bridges; and,

5. A statement of the improvements necessary to be made on such roads and bridges, and an estimate of the probable expense of making such improvements, beyond what the labor to be assessed in that year will accomplish.

See Appendix, No. 4.

Repairs of roads and bridges.

§ 4. The commissioners of highways of each town shall deliver to the supervisor of such town, a statement of the improvements necessary to be made on the roads and bridges, together with the probable expense thereof; which supervisor shall lay the same before the board of supervisors at their next meeting. The board of supervisors shall cause the amount so estimated, to be assessed, levied and collected, in such town, in the same manner as other town charges; but the moneys to be raised in any such town, shall not exceed in any one year, the sum of two hundred and fifty dollars.

See form No. 6.

In April, 1832, an act was passed "*for the more effectual improvement of Roads and Bridges.*" See Session Laws of 1832, p. 480, also, 1 R. S. 341.

§ 1. "Whenever the commissioners of highways of any town in this state, shall be of opinion, that the sum of two hundred and fifty dollars, as now allowed by law, will be insufficient to pay the expenses actually necessary for the improvement of roads and bridges, it shall be lawful for such commissioners to apply in open town meeting, for a vote authorising such additional sum to be raised as they may deem necessary for the purpose aforesaid, not exceeding two hundred and fifty dollars in addition to the sum now allowed by law." See Appendix, No. 11.

§ 2. "Before making such application, it shall be the duty of the commissioners to give notice of their intended application, by posting the same in a conspicuous manner, in at least five of the most public places in such town, at least four weeks next preceding the annual town meeting; such notice shall specify the amount to be applied for, and the purposes for which the same is intended to be appropriated, with the probable amount necessary to be expended at each place, if there shall be more than one."

§ 3. "Whenever an application for a grant of money for the purpose mentioned in the first section of this act, shall be made to any town meeting, it shall be the duty of the commissioners making the same, to exhibit a statement of their accounts, and an estimate of the expenses necessary for the improvement of roads and bridges, in such town the ensuing year."

§ 4. "If the town meeting shall, by their votes, determine that a sum over and above the amount now allowed by law, will be necessary for the improvement of roads and bridges, or to pay any balance that may be due; the clerk shall enter such resolution as shall be agreed to, in the minutes of the meeting, and deliver a copy thereof to the supervisors of the town, who shall lay the same before the board of supervisors, at their next annual meeting; and it shall be their duty to cause the amount specified in such resolution to be levied and collected, in the same manner as other town charges of such town."

A further act was passed April 18, 1838, entitled "an act to enlarge the powers of the board of supervisors," p. 314—by which the supervisors were authorized:

§ 1. Sub. 5. To cause to be levied, collected and paid in the manner now provided by law, such sum of money, in addition to the sum now allowed by law, not exceeding five hundred dollars in any one year, as a majority of the qualified voters of any town may at any legal town meeting have voted to be raised upon their town, for constructing or repairing roads and bridges in such town.

§ 2. No moneys shall be raised under the authority conferred by the fifth subdivision of the preceding section, unless a written notice of the application to such town meeting to raise such amount, shall be posted on the door of the house where the town meeting is to be held, and also at three public public places in such town two weeks before the town meeting, and be also openly read to the electors present, immediately after the opening of the meeting.

---

§ 5. It shall be the duty of commissioners of each town, to cause mile-boards or stones to be erected, where not already erected, on the post roads, and such other public roads in their town, as they may think proper, at the distance of one mile from each other, with such fair and legible inscriptions as they may think proper. Mile-stones.

This duty should be performed when the funds are suffcient in a town, but should not to the neglect of the roads and bridges.

OVERSEERS OF HIGHWAYS.

Overseers, their duty.

§ 6. It shall be the duty of the overseers of highways in each town,

1. To repair and keep in order the highways within the several districts for which they shall have been elected:

2. When so required by the commissioners of highways, or any one of them, to warn all persons assessed to work on the highways in their respective districts, to come and work thereon:

3. To cause the noxious weeds on each side of the highway within their respective districts, to be cut down or destroyed twice in each year, once before the first day of July, and again before the first day of September; and the requisite labor shall be considered highway work: and,

4. To collect all fines and commutation money, and to execute all lawful orders of the commissioners.

By § 1 R. S. 339, the town shall choose at the annual town meeting, as many overseers of highways as there are road districts in the town.

SUB. 1. The overseer is to diligently repair and work the road within his district to the amount of the rates assessed by the commissioners. He is not bound to do more.

1. A penalty is given for neglect. § 16.
2. He is liable to an indictment for a neglect of duty.
3. To an action at the suit of a party injured. An indictment will lie in pursuance of § 38, 2 R. S. 676, which declares that when any duty is or shall be enjoined by law upon any public officer, every wilful neglect to perform such duty, when no special provision shall have been made for the punishment of such delinquency, shall be punishable as a misdemeanor.

No action for private damages will lie against an overseer, for not *repairing a bridge.* This is distinctly settled in 17 Johns. 439.

"All the powers of the overseer must therefore be taken to be subordinate to, and under the superior control of the orders of the commissioners, whom they are bound to obey. It is further to be observed that the duty of overseers is confined to the highways, and it is the commissioners alone, who are directed to keep in repair bridges as well as highways. The overseers have no concern with bridges over streams, except so far as they are directed generally to execute the orders of the commissioners."

What shall be deemed a bridge is a question to be determined according to the ordinary meaning of the word. Culverts or sluices over small brooks or streams, and even bridges of some extent in a road district are under the care of the overseer and he is bound to repair them. See notes under § 1. page 9.

As to an action for personal damages for not repairing a highway; there is no doubt expressed by Chancellor Kent, in the case last quoted, but it was not the question before the court, and therefore, it is not settled.

Undoubtedly, an action can be maintained by the party injured, if he can show that the overseer had the means, and wilfully neglected, or refused to make the repairs.

In the case of *Adsit vs. Brady,* (4 *Hill,* 630,) the court say:— "when an individual sustains an injury by the misfeasance of a public officer, who acts or omits to act contrary to his duty; the law gives redress to the injured party by an action adapted to the nature of the case." This is quoted from the words of Spencer, 15, *Johns R.* 253. *Russel vs. Men of Devon,* (2 *Term. Rep.* 671.

SUB. 2. When the commissioners deliver the rate bill, the overseers, without any special directions, are bound to proceed with reasonable diligence, and warn the persons and have the work done. They violate their duty if they do not, and they must not wait for special directions from the commissioners.

SUB. 3. Undoubtedly the performance of this duty rests in the sound discretion of the overseer. If such weeds are so situated as it is not likely they will do injury; and in cases where the road is very bad and little work in the district, it would be absurd to expend it in destroying weeds instead of doing necessary work.

SUB. 4. *All Fines.* 1. For idleness, § 1. 2. For not working, § 42. Swinging gates, § 171. 4. He can also prosecute, under § 12, when a person assessed for a scraper or plough, refuses to pay.

All forfeitures and penalties not specially provided for, are to be sued for by the commissioners.

Some of the duties of overseers are to be performed without any special order or direction from the commissioners; it is their duty among other things, to keep the roads in repair; to destroy noxious weeds; to collect fine and commutation money, and to remove loose stones from the beaten track of the road, without any special order from the commissioners. It would seem from Sub. 2, that they are not to warn persons assessed to work without being required by the commissioners to do so; but this requisition relates to the general warrant directed to them by the commissioners.

And the court decided that an overseer is bound to remove obstructions from the highways within his district, although not specially directed by the commissioners. *McFadden vs. Kingsbury,* 11 *Wend.* 667.

In that case the court say that "by the 6th Section, it is made the duty of overseers to repair and keep in order the highways within his district. What is meant by keeping a road in order? If a fence is built across a road in any district, any person may remove it; but is it not the duty of the overseers to see that this is done? If logs are thrown in the road, or rubbish of any description placed in it, which deprive the public of the use of the road whose duty is it to remove those obstacles to the public enjoyment of the right of passage upon the road? It is undoubtedly the duty of the overseer. It is no answer to say that the commissioners have the care and superintendence of the highways, and give directions for the repairing of roads and bridges; it is not the less the duty of the overseers to repair and keep in order the road in their districts; they are bound to do it, whether they receive special instructions or not; it is their duty without special orders."

Overseers, their duty.

§ 7. It shall be the further duty of the overseers of highways once in every month, from the first day of April to the first day of December, to cause all the loose stones lying on the beaten track of every road within their respective districts, to be removed; and to cause the monuments erected or to be erected as the boundaries of highways to be kept up and renewed, so that the extent of such roads may be publicly known.

This is a duty very much neglected and if more punctually and faithfully performed would save much difficulty and many law suits.

New assessments to be made by them.

§ 8. When the quantity of labor assessed on the inhabitants of any road district by the commissioners, shall be deemed insufficient by the overseer of such district to keep the roads therein in repair, it shall be the further duty of such overseer, to make another assessment on the actual residents in such district, in the same proportion, as near as may be, and not exceeding one-third of the number of days assessed in the same year by the commissioners on the inhabitants of such district; and the labor so assessed by an overseer, shall be performed or commuted for, in like manner as if the same had been assessed by commissioners of highways.

The assessing of the third on the district is entirely in the discretion of the overseer. If the road is in a very bad and dangerous condition, he should do it. He is to exercise his best judgment, and an overseer is not liable for a mere error in judgment. 10 *Johns.* 470. *See Appendix, No.* 5.

§ 9. The commissioners of highways of each town shall cause guide-posts, with proper inscriptions and devices, to be erected at the intersections of all the post-roads in their town, and at the intersection of such other roads therein as they may deem necessary. Guide-posts.

§ 10. It shall be the duty of the overseers of highways of each town, to maintain and keep in repair, at the expense of the town, such guide-posts as may have been erected by order of the commissioners, within the limits of the districts, for which they shall have been respectively elected or appointed. Ibid.

§ 11. The commissioners of highways, whenever they shall think it necessary or useful, may direct and empower any overseer of highways in their respective towns, to procure a good and sufficient iron or steel-shod scraper, and plough, or either of them, for the use of his road district; to be paid for, by the moneys arising from commutations and fines within such district. Scrapers and plows.

§ 12. In case such moneys shall be insufficient for the purpose, the deficiency shall be assessed by the overseers upon the inhabitants of the districts, in the proportion they are respectively assessed, on the as- Ibid.

sessment roll of said town; and if any one so assessed shall neglect or refuse to pay such assessment, the same may be used for and recovered by the overseer.

This is an assessment in money. The overseer can either make a new assessment roll, or carry out the amount on the original assessment, opposite their names. If there is likely to be an opposition to the collection of the amount, the better way will be to make a new one. *See form Appendix, No.* 7.

Excess of work by Overseers.

§ 13. If any overseer shall be employed more days in executing the several duties enjoined on him by this Chapter, than he is assessed to work on the highway, he shall be paid for the excess at the rate of seventy-five cents per day, and be allowed to retain the same out the moneys which may come into his hands for fines under this Chapter; but he shall not be permitted to commute for the days he is assessed.

The overseer must work out all his assessment. And if to direct the men employed, he works more days, or if in executing other duties he works more than his time, he must take it out of moneys in his hands. But if no fines or commutation money is received by him, how is he to get his pay? There seems to be no provision made for his remuneration.

Vacancy in their office.

§ 14. If any person chosen to the office of overseer of highways, shall refuse to serve, or if his office shall become vacant, the commissioners of highways of the town, shall, by warrant under their hands, appoint some other person in his stead: and the overseer so appointed, shall have the same powers, be subject to the same orders, and liable to the same penalties, as overseers chosen at town meeting.

The person refusing to serve, cannot be appointed, as the words of the section are "some other person." The legislature considered the penalty as an equivalent for the service. If it were not so, a person refusing to serve for any urgent reasons might be subject to endless litigation. *Haywood vs. Wheeler*, 11 *Johns. R.* 432. *See form for appointment, Appendix, No.* 8.

§ 15. The commissioners making the appointment, shall cause such warrant to be forthwith filed in the office of the town clerk, who shall give notice to the person appointed, as in other cases. Proceedings.

§ 16. Every overseer of highways who shall refuse or neglect either, Penalties on overseers.

1. To warn the people assessed to work on the highways, when he shall have been required so to do, by the commissioners, or either of them;

2. To collect the moneys that may arise from fines or commutations; or,

3. To perform any of the duties required by this Chapter, or which may be enjoined on him by the commissioners of highways of his town, and for the omission of which, a penalty is not hereinafter provided;

Shall, for every such refusal or neglect, forfeit the sum of ten dollars, to be sued for by the commissioners of highways of the town; and when recovered, to be applied by them in making and improving the roads and bridges therein.

The delivery of the assessment roll is a sufficient requisition under the first and second subdivisions; he is also bound to destroy noxious weeds, to remove loose stones from the beaten track of the highway without special orders; but he is not obliged to procure scrapers, unless directed and empowered so to do. 11 *Wend.* 667.

§ 17. It shall be the duty of the commissioners of highways of each town, whenever any person resident in their town shall make complaint that any overseer of highways in such town has refused or neglected to perform any of the duties enumerated in the last preceding section, and shall give or offer to such commissioners, sufficient security to indem- To be prosecuted by commissioners.

nify them against the costs which may be incurred in prosecuting for the penalty annexed to such refusal or neglect, forthwith to prosecute such overseer for the offences complained of.

By the former highway act, half the penalty went to the complainant. The Revisers in their note under this section, say, that they "have uniformly omitted all provisions authorizing common informers to sue for their own benefit, and it is presumed the above "section will answer the purposes intended by the act of 1826."

It is the duty of the commissioners to sue without complaint. In general, all public officers, though not expressly authorized by statute, have a capacity to sue, commensurate with their public trusts and duties. (18 *Johns. R.* 407. 1 *Cow.* 260, *and Note, p.* 261—4. 3 *Wend.* 193. 7 *Wend.* 181. 19 *Wend.* 50.) The suit cannot be maintained in the official name or title, but must use their individual name, annexing their official title. *Commissioners of Courtlandville vs. Peck,* 5 *Hill,* 215. 4 *Hill,* 136. *See Appendix, Nos.* 9 & 10.

Penalty for neglect.

§ 18. If such commissioners of highways shall refuse or neglect to prosecute for such penalty, they shall, in every such case forfeit the sum of ten dollars, to be recovered by the person who shall have made such complaint, and given or offered such security.

ARTICLE SECOND.

Of the Persons liable to work on Highways, and the making of Assessment therefor.

SEC. 19. Who, and what property liable to be assessed for highway labor.
20. When commissioners of highways to meet.
21. Overseers to deliver list of persons liable to work on highway, to town clerk.
22. Non-resident lands how to be ascertained and appraised.
23. Town clerk to deliver lists received by him, to commissioners of highways.
24. Mode of proceeding to assess highway labor.
25. Monied corporations may be assessed.
26. Notice to be given; labor how and when performed.
27. May commute; money how expended.
28 & 29. Penalties for default; how collected.
30. Commissioners to assess property or persons omitted by assesors.
31. Powers of commissioners to be exercised by certain other officers in certain cases.
32 & 33. Copy of each list subscribed by commissioners, to be delivered to overseers.
34. Improved lands of non-residents to be assessed as if its owners were residents.

§ 19. Every person owning or occupying land in the town in which he or she resides, and every male inhabitant above the age of twenty-one years, residing in the town when the assessment is made, shall be assessed to work on the public highways in such town; and the lands of non-residents, situated in such town, shall be assessed for highway labor, as herein after directed. Persons liable to be assessed.

A question was raised in *Beach vs. Furman*, (*Johns. R.* 529,) whether a female, though a freeholder, was liable to a highway assessment. To obviate that question, this section was made explicit, by inserting the word *he* or *she*, by the legislature. *See Revisors' Notes on this section.* Females are liable to be assessed on their property, and males are liable to a tax *per capita.*

§ 20. The commissioners of highways of each town, shall meet within eighteen days after they shall be chosen, at the place of town meeting, on such day as they shall agree upon and afterwards at such other times and places as they shall think proper. Meetings of commissioners.

At this meeting they are to receive from the town clerk the list of taxable inhabitants, to enable them to make the assessment. They can adjourn or appoint subsequent meeting. This section is directory, and if they should omit a meeting within the eighteen days, according to the usual construction of such statutes, they could afterwards meet, and their proceedings would not be void. The public should not suffer through their default. They might be liable to punishment for their neglect of duty.

§ 21. Each of the overseers of highways shall deliver to the clerk of the town, within sixteen days after his election or appointment, a list subscribed by such overseer, of the names of all the inhabitants List of inhabitants.

in his road district, who are liable to work on the highways.

This duty should be carefully performed.

This list does not include non-residents. It would be well for the overseer, however, to add all such to his list, that the commissioners may have all the information on the subject. *See form, Appendix, No.* 12.

Non-residen ands how appraised.

§ 22. The commissioners of highways in each town, at their first or any subsequent meeting, shall make out a list and statement of the contents of all lots, pieces or parcels of land within such town, owned by non-residents therein, every lot so designated shall be described in the same manner as is required from assessors, and its value shall be set down opposite to such description, such value shall be the same as was affixed to such lot in the last assessment roll of the town; and if such lot was not separately valued in such roll, then in proportion to the valuation which shall have been affixed to the old tract of which such lot shall be a part.

See Appendix, No. 13, *for form of list and statement.*

The commissioners can have the use of the assessment roll deposited with the town clerk. The manner in which assessors are required to describe the lots will be found, 1 *R. S. p.* 391, § 11, 12, 13 and 14. It will be advisable to follow the assessment roll, except where the commissioners are bound to make a proportion, and then they should be very particular in their description of the proportions.

Lists of inhabitants.

§ 23. The town clerk shall deliver the lists filed by the overseers to the commissioners of highways of the town; who shall proceed, at their next meeting, or at some subsequent meeting to ascertain, estimate and assess the highway labor to be performed in their town, the then ensuing year.

When there is only one commissioner in a town, he should within eighteen days after election, repair to the place where the town meeting is held according to the §20, page 21, for the purpose of making out his lists. He should appear at that place to give all persons opportunity to object to the assessment, and to furnish any explanations or information in reference to the owners of property, or what changes have taken place since the last assessment.

§ 24. In making such estimate and assessment, the commissioners shall proceed as follows:

Proceedings in making assessments.

1. The whole number of days' work to be assessed in each year, shall be ascertained, and shall be at least three times the number of taxable inhabitants in such town;

2. Every male inhabitant being above the age of twenty-one years (excepting ministers of the gospel, and priests of every denomination, paupers, and idiots, and lunatics,) shall be assessed at least one day;

3. The residue of such days' work, shall be apportioned upon the estate real and personal of every inhabitant of such town, as the same shall appear by the last assessment roll of the said town, and upon each tract or parcel of land, which the owners are non-residents, contained in the lists made as aforessid;

4. If, after such apportionment, there shall be any deficiency in the number of days' work determined by the commissioners, to be performed in their town, the then ensuing year, such deficiency shall be assessed upon the estates real and personal of the inhabitants of the town, and upon each tract or parcel of land, of which the owners are non-residents, according to the last assessment roll;

5. The commissioners shall affix to the names of each person named in the list furnished by the overseers, and also to the description of each tract or parcel of land contained in the list prepared by them, of non-resident lands, the number of days which such

person or tract shall be assessed for highway labor, as herein directed, and the commissioners shall subscribe such list and file them with the town clerk.

SUB. 1. Three times the number of taxable inhabitants is the number of days' works to be assessed, unless they shall determine to add to that number. This they have the power to do, and there does not appear to be any limitation to their discretion as to the number which they may add.

SUB. 2. *Ministers,* or priests who officiate according to the rules of their denomination, who are engaged in religious teaching as a business; not those who have abandoned or retired from such calling; nor will it include those who have not been engaged in religious teaching as a profession, but only occasionally teach or exhort.

An idiot, or a natural fool, is one who has had no understanding from his nativity; a lunatic, is one who has had understanding but by disease, or other accident, has lost the use of his reason, but who has intervals of reason called lucid intervals. 1 *Blackstone's Com.* 303—4.

For form under this §, *See Appendix, No.* 14.

This section in the first edition of the R. S. contained seven subdivisions. The 4th then provided, that only one quarter of a days' labor should be assessed for $100 on lands of non-residents; and the 5th subdivisions provided that the commissioners should not assess at all, unless in their judgment the lands are enhanced by the highway.

It was decided under section 19, in 1834. *Bank of Ithaca vs. King,* (12 *Wend.* 390,) that monied corporations were not liable to this assessment, on the ground they were not *persons.* A corporation having no material existence is incapable of performing labor. Although the word persons would, if alone and unexplained, include corporations, were the other provisions of the Statute consistent with such constructions, yet where the legislature use the words *he* or *she,* it is explained what kind of persons are intended. To render them liable for a highway assessment, the following act was passed May 15, 1837, p. 488.

Stock to be assessed.

§ 25. "When by law the powers and duties of commissioners of highways are conferred and imposed upon other officers, they shall possess all the powers and perform all the duties in this act conferred or enjoined upon work upon each of the male inhabitants therein, above the age of twenty-one years, as provided in the sixteenth chapter of the first part of the Revised Statutes, entitled "Of highways and bridges," the commissioners of highways shall include among the inhabitants of such town, among whom such residue is to be apportioned, all moneyed or stock corporations, which shall appear on the last assessment roll of their town to have been assessed therein." [*Sec.* 1, *of Chap.* 431, *of* 1837.

It was decided in *Supervisors of Niagara vs. The People,* 7, *Hill* 504, that associations formed under the general banking law are corporations within the meaning of 1 R. S. 414, § 1, and liable to taxation on their capital. *See also* 4 *Hill* 20.

§ 26. "Such corporations shall be notified to furnish the amount of highway labor assessed to them in the same manner as individuals residing in such town, by giving oral or written notice to the president, cashier, agents, treasurer, or secretary of such corporation, or any clerk or other officer thereof, at the principal office or place of transacting the business or concerns of the said company: which labor shall be performed in such district or districts as the commissioners of highways of the town shall direct, and any number of day's work, not exceeding fifty, may be required to be performed, by such corporation, in any one day." [*Sec.* 2, *of same chap.*] Notice to be given to corporations.

§ 27. "Every such corporation may commute for the highway labor assessed upon it, in the same manner and at the same rate as is allowed by law to individuals, or by paying such commutation to a commissioner of highways of the town: and the commutation money so paid may be expended by the commissioners of highways upon any district or districts in the town; and for that purpose the said commissioners shall be entitled to demand and receive from the overseers, to whom any such commutation may Corporations may commute.

have been paid, the whole or any portion thereof; but in every case where any such corporation shall be located in any city, village or town, where by the law the road tax is now payable in money, the road tax imposed on any such corporation shall be paid in money, according to the provisions of the several laws affecting said city, village or town." [*Sec.* 3. *same.*

Penalties. § 28. "Such corporations shall be liable to the same penalties for every days' work required, and for every default of any substitute sent by them, as is provided by law in the case of individuals required to work on highways, which shall be collected in the same manner, and paid over to the commissioners of highways of the town, by the constable collecting the same, and may be expended by them in the same manner as herein provided for the commutation money received from any such corporation. The summons issued by any justice according to this act, may be for any number of penalties incurred by any such corporation previous thereto, and may be served in the manner provided by law for the service of writs or summons issuing out of courts of record against corporations." [*Sec.* 4, *same chap.*

How to be collected. § 29. "In case any such penalty cannot be collected as herein provided, the commissioners of highways of the town may file a bill in the court of chancery against any such delinquent corporation for the discovery and sequestration of its property; whereupon the same proceedings shall be had as are provided by law for the collection of county taxes assessed against incorporated companies; and the chancellor shall possess the like powers in respect to the same:

and the said commissioners may also recover such penalties, or any number of them that may have been incurred, with costs, from such delinquent company, from any court of record in this state." [*Sec.* 5, *same chap.*

Omisssins how to be rectified.

§ 30. "Whenever the assessors of any town shall have omitted to assess any inhabitant or property in such town, the commissioners of highways shall assess the persons and property so omitted, and shall apportion highway labor upon such persons or property in the same manner as if they had been duly assessed upon the last assessment roll." [*Sec.* 6, *same chap.*

Powers of commissioners of highways.

§ 31. "When by law the powers and duties of commissioners of highways are conferred and imposed upon other officers, they shall possess all the powers and perform all the duties in this act conferred or enjoined upon such commissioners, and the assessments under it for the present year shall be made before the first day of July next, by the commissioners of highways." [*Sec.* 7, *of same chap.*

Non-residents.

§ 32. Lands of non-residents within any town, occupied and improved by the owner or owners, or his or their servants or agents, shall be liable to the same assessments for highways as if the owner or owners were residents. [*Sec.* 1, *of chap.* 107, *of* 1832.

This section was added to explain the 24th section as it originally stood in the first edition; it is now nearly or quite useless, as by the 24th section all lands are to be assessed equally. The land is to be assessed according to its value, and the servant or agent is to be assessed his poll tax, and for his own private property, if he is assessed for any on the last year's assessment roll.

Land occupied by a servant.

§ 33. The real property of non-resident owners, improved or occupied by a servant or agent, shall be subject to assessment of highway labor, and at the same rate as the real property of resident owners. [*Sec.* 1, *of chap.* 154, 1835.

This section was also enacted to modify the original § 24, as it stood in the first edition, as the § 33 now reads, this section has no force.

Copies of lists.

§ 34. [*Sec.* 25.*] The commissioners shall direct the clerk of the town to make a copy of each list, and shall subscribe such copies; after which, they shall cause the several copies to be delivered to the respective overseers of highways of the several districts in which the highway labor is assessed.

Names Omitted.

§ 35. [*Sec.* 26.] The names of persons left out of any such list, and of new inhabitants, shall from time to time be added to the several lists, and they shall be rated by the overseers in proportion to their real and personal estate, to work on the highways, as others rated by the commissioners on such list, subject to an appeal to the commissioners.

Perhaps the overseer in making his list for the town clerk, or commissioners in making assessment, omitted names through inadvertance, or if new inhabitants have moved into the district, the overseers must put down such name on the bottom of his list, and assess them, or annex an assessment, as *No.* 15 *in Appendix.* He should notify them of their assessment immediately, that they may have an opportunity to appeal to the commissioners.

No particular form is necessary to be used by overseers. *See Appendix, No.* 16.

The appeal to commissioners need not be in writing. It will be safer for the appellant to make it in writing, from a form which will be found in *Appendix No.* 17. Serve a copy of this on the commissioners, and also on the overseer, and let the commissioners appoint a day and place for hearing, and notify the overseer. If the commissioners are of opinion that the appellant is not taxable, they will strike his name from the list, or they can reduce the number of days if they decide that he is assessed for too many.

Appeals by non-residents.

§ 36. [*Sec.* 27.] Whenever any non-resident owner shall conceive himself aggrieved by the assessment of

* The second numbers in brackets refer to the first ed. of R. S.; the others to the third ed.

any commissioners of highways, in carrying into effect the provisions of this article, it shall be lawful for such owner, or his agent, within thirty days after such assessment, to appeal to any three judges of the court of common pleas, of the county in which such such land is situated.

This appeal must be served within thirty days after the assessment is actually made, and the list signed by the commissioners. Notice to be served on such commissioners.

Let the judge appoint the time and place for hearing, and serve notice on commissioners.

§ 37. [*Sec.* 28] It shall be the duty of such judges, within twenty days thereafter, to convene and decide on such appeal, the said owner or agent giving notice to the commissioners of the time of the meeting of the judges; and their decision, or that of any two of them, shall be final and conclusive in the premises. Each judge shall be entitled to receive for services on such appeal, two dollars for each day he may be employed thereon, to be paid by the party appealing, if the proceedings of commissioners and overseers shall be affirmed; but if reversed or modified favorable to the party appealing, to be levied and paid as part of contingent expenses of such town. Proceedings.

As there are now by the new constitution, no judges of the court of common pleas, the two last sections appear to be inoperative.

§ 38. [*Sec.* 29.] It shall be the duty of the commissioners of highways of each town, to credit such persons as live on private roads, work the same, so much on account of their assessments as such commissioners may deem necessary to work such private roads, or to annex such private roads to some of the highway districts. Private roads.

§ 39. [*Sec.* 30.] Whenever the commissioners of highways shall assess the occupant, for any land not Certain assessments to be separate.

owned by such occupant, they shall distinguish in their assessment lists, the amount charged upon such land, from the personal tax, if any, of the occupant thereof. But when any such land shall be assessed in the name of the occupant, the owner thereof shall not be assessed during the same year to work on the highways on account of the same land.

Tenant to deduct assessment.

§ 40. [*Sec.* 31.] Whenever any tenant of any land for a less term than twenty-five years, shall be assessed to work on the highways, for such land, pursuant to the last preceding section, and shall actually perform such work, or commute therefor, he shall be entitled to a deduction from the rent due, or to become due from him, for such land, equal to the full amount of such assessment, estimating the same at the rate of sixty-two and a half cents per day; unless otherwise provided for covenant or agreement, between such tenant and his landlord.

ARTICLE THIRD.

On the Duties of Overseers in regard to the Performance of Labor upon Highways; and of the Performance of such Labor or the Commutation therefor.

SEC. 41. Overseers to give twenty-four hours' notice to persons assessed to work on highways.
42. Notice when to be given to agents of non-residents.
43. How given where no agent in the town.
44. Persons to work their whole number of days, unless they commute.
45. Persons commuting, to pay in twenty-four hours after notice, to appear and work.
46. Overseers may require a team, &c., from persons assessed for three days.
47. Persons may appear by substitute; hours to work.
48. Persons not working faithfully to forfeit one dollar.
49. Penalty for not commuting or not appearing, &c.
50. Overseers to make complaint of delinquents to a justice.
51. Justice to summon delinquent to show cause why he should not be fined.
52. If not sufficient cause be shown, delinquent to be fined; warrant to be issued.
53. Money collected, to go to overseer who entered complaint; how applied.
54. Penalties collected to be set off against assessment.
55. Acceptance of excuse by overseer, not to exempt from commuting or working.

§ 41. [*Sec.* 32.] It shall be the duty of the overseers of highways, to give at least twenty-four hours' notice to all persons assessed to work on the highways, and residing within the limits of their respective districts, of the time and place, when and where they are to appear for that purpose, and with what implements; but no person, being a resident of the town, shall be required to work on any highway, other than in the district in which he resides, unless he shall elect to work in some district where he has any land; and in such case he may, with the approbation of the commissioners of highways, apply the work assessed in respect to such land, in the district where the same is situated. Notice to work. Where to be done.

It requires no formal application to the commissioners to obtain their consent to work in another district, nor need they express their approbation in any formal manner; but to avoid any controversy, it will be well in such cases to obtain this written consent. *See Appendix, No.* 19.

Let the applicant work the number of days, and bring a receipt or certificate from that overseer to the one where he resides, who will give him credit accordingly.

§ 42. [*Sec.* 33.] It shall be the duty of the several overseers of highways, to notify the agent of every non-resident landholder, whose lands are assessed, (if such agent reside in the town where such assessment is made,) of the number of days such non-resident is assessed, and of the time when, and the place where, Notice to non-residents.

the labor is to performed; which notice shall be given at least five days previous to the time appointed.

It is not necessary that this notice should be in writing; but as a matter of prudence it may be well to serve a written notice; the agent may forget, and may not inform his principal; and as he can be a witness, the overseer had better serve notice and take from him an admission of service. *See Appendix, No.* 20.

Ibid. § 43. [*Sec.* 34.] If the overseer cannot ascertain that such non-resident has an agent within such town, he shall affix a written notice on the outer door of the building in which the last town meeting in such town was held, containing a list of the names of such non-residents, when known, and a description of the tracts of land comprised in his list, together with the number of days' labor assessed on each tract, and a specification of the time when, and place where, such labor is to be performed; which notice shall be posted at least twenty days before the time appointed for performing such labor.

Form, See Appendix, No. 21.

§ 44. [*Sec.* 35.] Every person liable to work on the highways, shall work the whole number of days for which he shall have been assessed; but every such person, other than an overseer, may elect to commute for the same, or for some part thereof, at the rate of sixty-two and a half cents for each day; in which case such commutation money shall be paid to the overseer of highways of the district in which the person commuting shall reside, to be applied and expended by such overseer in the improvement in the roads and bridges in the same district.

Deductions are to be made according to the following Statute:— "Every non-commissioned officer and musician who shall produce

"to the overseer of highways, or person authorized to receive com-
"mutations for highway labor, a certificate from the commandant of
"his company, of his being qualified, and having done military duty,
"as required by law, for the preceding year, shall be entitled to a
"deduction from his labor on the highways, or from his commuta-
"tion for such labor, two days." 1 *R. S.* 324, § 4.

The overseer is required to work out his tax personally, that he may be present, work well, and superintend the labor of others.

§ 45. [*Sec.* 36.] Every person intending to commute for his assessment, or for any part thereof, shall, within twenty-four hours after he shall be notified to appear and work on the highways, pay the commutation money for the work required of him by such notice; and the commutation shall not be considered as complete until such money be paid. When to be paid.

§ 46. [*Sec.* 37.] Every overseer of highways shall have power to require a team, or a cart, wagon, or plough, with a pair of horses or oxen, and a man to manage them, from any person having the same within his district, who shall have been assessed three days or more, and who shall not have commuted for his assessment; and the person furnishing the same upon such requisition, shall be entitled to a credit of three days for each days' service therewith. Teams, &c.

§ 47. [*Sec.* 38.] Every person assessed to work on the highways, and warned to work, may appear in person, or by an able bodied man, as a substitute; and the person or substitute so appearing, shall actually work eight hours in each day, under the penalty of twelve and a half cents for every hour such person or substitute shall be in default, to be imposed as a fine on the person assessed. Substitutes, hours to work.

§ 48. [*Sec.* 39.] If any such person, or his substitute, shall, after appearing, remain idle, or not work faithfully, or hinder others from working, such of- Penalty for neglect, &c.

fender shall, for every offence, forfeit the sum of one dollar.

Penalties for not working. § 49. [*Sec.* 40.] Every person so assessed and duly notified, who shall not commute, and who shall refuse or neglect to appear as above provided, shall forfeit for every days' refusal or neglect, the sum of one dollar. If he was required to furnish a team, carriage, man, or implements, and shall refuse or neglect to comply, he shall be fined as follows:

1. For wholly omitting to comply with such requisition, three dollars for each day;

2. For omitting to furnish a cart, wagon, or plough, one dollar for each day;

3. For omitting to furnish a pair of horses or oxen, one dollar for each day.

4. For omitting to furnish a man to manage the team, one dollar for each day.

Complaints, how made. § 50. [*Sec.* 41.] It shall be the duty of every overseer of highways, within six days after any person so assessed and notified, shall be guilty of any refusal or neglect for which a penalty or fine is prescribed in this Title, unless satisfactory excuse shall be rendered to him for such refusal or neglect, to make complaint on oath, to one of the justices of the peace of the town.

This section confers the power on the overseers of judging of the sufficiency of the delinquent's excuse. The party named may have a good and satisfactory excuse, as sickness, &c. Of this the overseer is judge. If he should decide that the excuse is not satisfactory, and should enter the complaint even unreasonably, he is not liable to the party assessed, unless he acts *maliciously,* 1 *Johns. R.* 515. 10 *Johns R.* 470. Nor is the overseer answerable, in private action, for any error of judgment in the execution of his trust. *See* 3 *Johns.* 474. If he acts maliciously, it is otherwise. 5 *Johns. R.* 125. Complaints must be made to the justice of the same town. *See form of complaint in Appendix, Nos.* 22 *and* 23.

In a summary proceeding on the complaint of an overseer of highways to fine a person assessed to perform highway labor, for neglect to attend when notified, the question whether the complainant is an overseer of highways of the town, is a jurisdictional fact. It may therefore be controverted when the warrant for a fine is offered in evidence in a collateral suit, *e. g.* in trespass by a party fined against the party on whose complaint the fine is imposed. *Walker vs. Mosely*, 5, *Denio* 102.

§ 51. [*Sec.* 42.] The justice to whom such complaint shall be made, shall forthwith issue a summons directed to any constable of the town, requiring him to summon such delinquent to appear forthwith before such justice, at some place to be specified in the summons, to show cause why he should not be fined according to law for such refusal or neglect; which summons shall be served personally, or by leaving a copy at his personal abode. Proceedings.

See form for Summons, Appendix, No. 31.

It must be served by a constable of the same town. These proceedings are very summary. The justice is authorized to proceed on the return of summons served by copy; the constable should therefore, be careful and make a personal service, if in his power. *See constable's return, Appendix, Nos.* 24 *and* 26.

§ 52. [*Sec.* 43.] If, upon the return of such summons, no sufficient cause shall be shown to the contrary, the justice shall impose such fine as is provided in this title for the offence complained of, and shall forthwith issue a warrant under his hand and seal, directed to any constable of the town where such delinquent shall reside, commanding him to levy such fine, with the costs of their proceedings, of the goods and chattles of such delinquent. Ibid.

When the summons is returned, the justice should allow a reasonable time for the party to appear and defend; and when it is returned, served by copy, he should see that all has been fair in the attempt to serve it. No adjournment can be granted; but the justice can exercise a reasonable discretion in holding open the court for the appearance of the delinquent, or to allow him to procure witnesses. The justice has a large discretion in passing over the sufficiency of the excuse. He is the sole judge, and his decision is final. No appeal or certiorari will lie to his decision.

By the former highway acts, the complaint of the overseer was final and conclusive, and the justice had no discretion, and acted merely as a ministerial officer. *See* 9 *Johns. R.* 229. 3 *Johns. R.* 474. 2 *Cain*, 179. The overseer can adjudge who was in default, and demand the warrant, and although the complaint is unjust, no action lay against the justice. *See form for Warrant, Nos.* 25 *and* 27.

Proceedings.

§ 53. [*Sec.* 44.] The constable to whom such warrant shall be directed, shall forthwith collect the moneys therein mentioned. He shall pay the fine, when collected, to the justice who issued the warrant, who is hereby required to pay the same to the overseer who entered the complaint, and to be by him expended in improving the roads and bridges in the district of which he is overseer.

It is questionable when any property is exempt from a levy under this warrant. The exemption is "under any execution" in the justice's act, and the same words were used in reference to executions issued out of courts of record, 2 *R. S.* 367 § 22. There has been no judicial exposition of the subject. We think, however, that the exemption does not apply to this case.

Penalties to be set off.

§ 54. [*Sec.* 45] Every penalty collected for a refusal or a neglect to appear and work on the highways, shall be set off against the assessment upon which it was founded, estimating every dollar collected as a satisfaction for one days' work.

Excuses.

§ 55. [*Sec.* 46.] The acceptance by an overseer of any excuse for refusal or neglect, shall not, in any case, exempt the person excused from commuting for or working the whole number of days for which he shall have been assessed during the year.

If the overseer, by excusing the person, could release him from the tax, he might defeat the object of the tax, by favoritism releasing those whom he pleases from performing any highway labor at all.

§ 56. [*Sec.* 47.] Every overseer of highways shall, on or before the first day of October, in each year, make out and deliver to the supervisor of his town, a list of all the lands of non-residents, and of persons unknown, which were taxed on his lists, on which the labor assessed by the commissioners of highways has been paid, and the amount of labor unpaid; and the said overseer, previous to delivering such list, shall make and subscribe an affidavit thereon, before some justice of the peace of such town, that he has given the notice required by the thirty-second (32nd) and thirty-third (33d) sections of this Title, and that the labor for which such land is returned, has not been performed. Proceedings to collect non resident labor unpaid.

See form in Appendix, Nos. 28, 29.

§ 57. [*Sec.* 48.] If any overseer shall refuse or neglect to deliver such list to the supervisor, as provided in the last preceding section, or shall refuse or neglect to make the affidavit as therein directed, he shall, for every such offence, forfeit the sum of five dollars, and also the amount of tax or taxes for labor remaining unpaid, at the rate of sixty-two and a half cents for each day; to be recovered by the commissioners of highways of the town, and to be applied by them in making and improving the roads and bridges in such town. Proceedings, &c.

§ 58. [*Sec.* 49.] It shall be the duty of the supervisors of the several towns, to receive the lists of the overseers of highways, when delivered pursuant to the preceding forty-seventh (47th) section, and to lay the same before the board of supervisors of the county. Ibid.

§ 59. [*Sec.* 50.] It shall be the duty of such Ibid.

board, at their next meeting, to cause the amount of such arrearages of labor, (estimating a day's labor at sixty-two and a half cents,) to be levied on the lands so returned, and to be collected in the same manner that the contingent charges of the county are levied and collected, and to order the same when collected, to be paid over to the commissioners of highways of the town, to be by them applied to the construction and improvement of the roads and bridges in the district, for whose benefit the labor was originally assessed.

Annual return of overseers. § 60. [*Sec.* 51.] Every overseer of highways shall, on the second Tuesday next preceding the time of holding the annual town meeting in his town, within the year for which he is elected or appointed, render to one of the commissioners of highways of the town, an account in writing, verified by his oath, (which the commissioners of highways are authorized to administer,*) and containing:

1. The names of all persons assessed to work on the highways, in the district of which he is overseer;

2. The names of all those who have actually worked on the highways, with the number of days they have so worked;

3. The names of all those who have been fined, and the sums in which they have been fined;

4. The names of all those who have commuted, and the manner in which the moneys arising from fines and commutations have been expended by him;

5. A list of all lands which he has returned to the supervisor for non-payment of taxes, and the amount of tax on each tract of land so returned.

See form in Appendix, No. 30.

* Laws 1833, Chap. 149.

§ 61. [*Sec.* 52.] Every such overseer shall also then and there pay to the commissioners, all moneys remaining in his hands unexpended, to be applied by the commissioners in making and improving the roads and bridges in the town, in such manner as they shall direct.

To pay over moneys.

§ 62. [*Sec.* 53.] If any overseer shall refuse or neglect to render such account, or having rendered the same, he shall refuse or neglect to pay any balance which may then be due from him, he shall for every such offence forfeit the sum of five dollars, to be recovered, with the balance of moneys remaining in his hands, by the commissioners of highways of the town, and to be applied in making and improving the roads and bridges. It shall be the duty of the commissioners of highways to prosecute for such penalty in every instance in which no return is made.

Penalty, how collected.

§ 63. Whenever it shall appear from the annual return of any overseer of highways, made in pursuance of the fifty-first section of the sixteenth Chapter of Title first of the First Part of the Revised Statutes, that any person who was assessed to work on the highways, (other than non-residents,) has neglected to work the whole number of days to him assessed, and has not commuted for, or otherwise satisfied such deficiency, then it shall be the duty of the commissioners of highways to reassess such deficiency to the person so delinquent, at the next assessment of work for highway purposes, and to add to it his annual assessment.

Reassessment in case of neglect.

§ 64. Such reassessment shall not exonerate any overseer of highways from any penalty which he may

Overseers.

have incurred under the sixteenth section of the last aforesaid Chapter.

ARTICLE FOURTH.

Of the laying out of Public and Private Roads, and of the Alteration or discontinuance thereof.

By § 5 of chap. 314 of 1838, the Boards of Supervisors are authorized to appoint special commissioners to lay out roads.

SEC. 65 & 66. Who may apply to commissioners for alteration or discontinuance of roads.
67. When commissioners shall lay out, &c., road, survey to be made thereof.
68. Town clerk to put up copy of order of commissioners laying out, &c., road.
69. Roads not to be laid through gardens, &c., without consent of owner.
70. No highway to be laid out through enclosed land, unless certified to be necessary.
71. Persons applying for road through any such land, to cause notices to be posted.
72. Freeholders to be sworn to examine as to necessity of road.
73. If they deem road to be necessary to make certificate thereof.
74. No compensation to be allowed them.
75. Before commissioners determine to lay out road, to give notice to occupant.
76. They shall meet, and if road laid out, description of it to be made and filed.
77. When damages may be fixed by agreement. Road not to be opened until damages are assessed.
78 to 85. Damages how assessed, by a jury; summons for jury; proceedings.
86. Damages and expenses, how collected.
87. When value of a discontinued road to be deducted from damage of new.
88. When commissioners of different towns disagree, to meet together, &c.
89. When highway necessary upon line of two towns, by whom laid out.
90. Such highway to be divided into two or more road districts.
91. Each district to belong to the town to which allotted.
92. Provision as to all such highways heretofore laid out.
93. When private road applied for, twelve freeholders to be appointed to examine.
94. If they determine road to be necessary, to make certificate thereof; damages, &c.
95. By whom and for what purpose road used.
96. Public roads not to be less, and private not more, than three rods wide.
97 & 98. Applications to discontinue an old road; proceedings thereon.
99. All papers relating to laying out, &c., road, to be filed in town clerk's office.
100. Persons aggrieved, may appeal to three common pleas judges.
101. Judges to whom first appeal made, to have exclusive jurisdiction.
102 to 105. Appeal to be in writing; its contents; proceedings thereon.
106. Each judge entitled to two dollars a day; by whom paid.
107. If office of a judge become vacant, another to be associated.
108 Road fixed by judges, not to be altered, &c., without their order.
109. If no one of the judges be in commission, application to whom to be made.
110. Application under either of two last sections to be accompanied by certificate.
111. When commissioners to give owners notice to remove fences.

§ 65. [*Sec.* 54.] Every person liable to be assessed for highway labor, [and every person owning lands in a town in which he is not a resident,] may apply to the commissioners of highways of the town in which he shall reside, to alter or discontinue any road, or to lay out any new road. Every such application shall be in writing, addressed to the commissioners, and signed by the person applying. Who may apply.

In all cases where the application is to lay out a road through enclosed, improved or cultivated lands, the description of the intended road should be particular, as it is the basis of the subsequent proceedings in calling out the freeholders. *See Notes under* § 57, 58 *and* 58. *See form, Appendix, Nos.* 32 *and* 33.

To give the commissioners *jurisdiction* the application must be made to them in *writing* by a person liable to be assessed for highway labor. *Harrington vs. the People*, 6 *Barb.* 611.

In that case there was an appeal for the decision of the commissioners refusing to lay out a road—and the judges of Washington county laid the road, but the order of the judges did not state that an application had been made in pursuance of this section—and the defendant in that suit offered to show on trial, that such written application had been made to the commissioners. The court say : "If "the commissioners had no jurisdiction of the application to lay out "the road, their determination was void; and the appeal from that "determination and the decision on the appeal was also void. If no "legal foundation was laid for the appeal, both appeal, and all the "proceedings therein were nullities." 6 *Barbour, S. C.* 611.

§ 66. Every person liable to be assessed for highway labor, and owning lands in a town in which he is not a resident, may apply to the commissioners of highways of the town in which the lands are situated, to alter, discontinue, or to lay out any road through the same. [*Chap.* 122 *of* 1836.]

Survey. § 67. [*Sec.* 55.] When the commissioners of highways shall lay out, alter, or discontinue any road, either upon application to them or otherwise, they shall cause a survey to be made of such road, and shall incorporate such survey in an order to be signed by them, and to be filed and recorded in the office of town clerk, who shall note the time of recording the same.

The commissioners have power to lay out a road through unenclosed, uncultivated or unimproved lands, without any application, in their judgment, whenever they shall decide it to be expedient.

Order to be posted. § 68. [*Sec.* 56.] It shall be the duty of the town clerk, whenever any order of the commissioners for laying out, altering, or discontinuing a road shall be received by him, to post a copy of such order on the door of the house where the town meeting is usually held: and the time hereinafter limited for appealing from any such order, shall be computed from the time of recording the same.

See Appendix, No. 34.

Consent of owner when necessary. § 69. [*Sec.* 57.] No public or private road shall be laid out through any orchard or garden, without the consent of the owner thereof, if such orchard be of the growth of four years or more, or if such garden have been cultivated four years or more, before the laying out of such road. Nor shall any such road be laid out through any buildings; or any fixtures or erections for the purposes of trade or manufactures; or any yards or enclosures necessary to the use and enjoyment thereof; without the consent of the owner.

These restrictions on the powers of the commissioners, apply only to cases where the owner withholds his consent. The commissioners may lay out a road through any kind of property with the consent of the owner, 4 *Paige*, 523. In the five cases enumerated, the commissioners have no power to lay out a road;—1. Through orchards; 2. gardens; 3. buildings; 4. fixtures and erections for trade, &c., and 5. yards and enclosures necessary for use of buildings and manufactures.

1. Orchard.—To be within the prohibition, it must injure the usefulness of the premises as an orchard, by taking some part where the fruit trees grow.

In *The People ex rel. Seward vs. the Judges of Dutchess*, 23 *Wend.* 360. The road was laid through a lane where two apple trees stood, that had formerly belonged to the orchard, but they had been separated from it by a fence for several years. The road also, as laid out, passed over the circular corner of a lot or field in which there was an orchard, including in the road a piece of ground about fifty feet long, and eight feet wide; but no apple tree in the orchard situated near the road. The court say, the road could not be laid out in such a manner as to deprive the owner, either whole or in part, of the beneficial enjoyment of his fruit trees. But we are, in effect asked to go further, and say that a road cannot be laid over an enclosed field, if there are fruit trees in any part of it, however distant they may be from the highway. To this doctrine we cannot subscribe. It does not follow that the whole field is an orchard because there are fruit trees in some part of it.

2. Garden.—It must have been cultivated as a garden four years previous to laying out. What constitutes a garden is a question of fact. The general definition, is a plat of ground appropriated to the cultivation of herbs or plants, fruits and flowers.

3. Buildings.—They must be buildings erected previous to the application and notice for the highway. If after such application and notice, the owner of the land puts buildings or erections thereon, it will not prevent the commissioners from going on and laying out the highway. It is the fault of the party himself, who cannot thus defeat the operation of the law. 2 *Hill*, 443.

Fixtures or Erections.—Only such as are for the purpose of trade and manufactures. Over other fixtures and erections, by the oath of freeholders, a road may be laid, if the erections do not amount to buildings. Tenter-bars erected for the purpose of carrying on the business of a fulling mill are within the prohibition. *Clark vs. Phelps*, 4 *Cow.* 190.

Neither a public nor a private road or way can be laid out across the fixtures and erections upon the inclined plane of a rail-road which are used for the drawing or letting down cars for the conveyance of merchandise or passengers. 6 *Paige*, 83.

What are or are not fixtures and erections, for the purposes mentioned in this section, is a question of fact, and must be judged of in reference to the situation and nature of the property.

5. YARDS OR ENCLOSURES.—In *ex parte. Clapp*, (3 *Hill.* 458,) the highway, as laid out, passed through the *door-yard* of one man, and left his *well, cow-shed*, and a part of his *corn-crib* in the highway. It also encroached on the *garden* and *cow-shed* of another man. It was decided by the court, that the commissioners had exceeded their powers, no consent being given. And they further decide, that the words *yards* or *enclosures* apply to "buildings" as well as to "fixtures or erections for the purpose of trade."

It is not every court-yard or enclosure which is appurtenant or contiguous to a dwelling-house or a manufacturing establishment, through which the commissioners are prohibited from laying out a road or highway. It is only such yards or enclosures as are necessary to the use and enjoyment of the dwelling-house or manufacturing establishment. This clause must be construed in reference to the situation and nature of the property to which the yard or enclosure is appurtenent. *Lansing vs. Caswell*, 4 *Paige*, 523.

The commissioners, as well as the judges, in laying out highways, exercise a special and limited jurisdiction; and although it may be presumed, until the contrary appear, that they have acted legally, it is quite clear that their acts may be impeached by showing that they exceeded their powers. 3 *Hill*, 461.

The commissioners should always take the care, when they attempt to lay a road by the permission of the owner, that his consent is given; and they should for the sake of caution, require it be in writing. For, if a man means, honestly, to give his consent, he will not hesitate to put it in writing.

The objection may be waived by the acts and conduct of the owner. Thus if he should appear before the judges on appeal, and litigate the question of damages, and appeal to the board of supervisors, it would undoubtedly be a legal consent to the laying of the road. 4 *Paige*, 523.

Oath, &c. when necessary.

§ 70. [*Sec.* 58.] No highway shall be laid out through enclosed, improved or cultivated land, without the consent of the owner or occupant thereof, unless certified to be necessary by the oath of twelve reputable freeholders of the town, in the manner hereinafter provided.

See Appendix, No. 35.

It is no objection that the certificate is signed by more thrn twelve freeholders; and in a case where the certificate was signed by twenty freeholders, the fact that five of them were of kin to the owner of the land, it was held that it did not vitiate it, there being still twelve who were properly qualified to sign. 7 *Wend.* 264.

Many of the remarks as to consent of owners, made under the last section, apply equally to this. A written consent is not necessary. The public acquire an interest in such road. Although there be no writing to pass the title of an owner, the Statute regulating highways provides for the laying out of highways by the consent of the owner. 6 *Wend.* 461.

A promise to pay the owner of lands a specific sum, on his consenting to have a public road laid through his lands, is not, for this reason, within the *Statute of frauds*, and may be enforced by an action, if such road be laid out and occupied as such. 6 *Wend.* 461.

This section uses the words owner or *occupant.* This does not mean that the consent of a tenant only can confer the right to take the premises upon the commissioners. If the occupant has an interest in the lands, his consent, as well as the owner's must be obtained. This is thought to be the correct view of the section, although there is no adjudication of the subject.

§ 71. [*Sec.* 59.] Every person who shall apply for the laying out of a highway through any such land, shall cause notices in writing to be posted up at three of the most public places of the town, specifying, as near as may be, the route of the proposed highway, the several tracts of land through which the same is proposed to be laid, and the time and place at which the freeholders will meet to examine the ground. Every such notice shall be posted up at least six days before the time specified therein for the meeting of the freeholders. Notice of application

See Appendix, No. 36.

If the *termini* and general route of the proposed road are given, without specifying the course and distances, it will be a sufficient compliance with the Statute. That is the business of the commissioners or judges if they conclude to lay out the road. 23 *Wend.* 360. *The People vs. Judges of Dutchess.*

§ 72. [*Sec.* 60.] If twelve reputable freeholders of the town, not interested in the lands through which the road is to be laid, nor of kin to the owner thereof, shall appear at the time and place specified in the notice, they shall be sworn by a justice of the Proceedings.

peace,* or any officer authorized to administer oaths, well and truly to examine and certify, in regard to the necessity and propriety of the highway applied for.

The commissioners have no authority to act, until *twelve* freeholders certify to the necessity and propriety of the road; but when *twenty* unite in a representation, *twelve* of course concur. In such case, if *five* were unqualified, still enough remained to make the number *twelve* who were qualified. 7 *Wend.* 264.

Ibid.

§ 73. [*Sec.* 61.] They shall then personally examine the route of such highway, and shall hear any reasons that may be offered for or against laying out the same. If they shall be of opinion that such highway is necessary and proper, they shall make and subscribe a certificate in writing to that effect, which shall be delivered to the commissioners of highways of the town.

See Appendix No. 37.

This section is very clear.

The jury decide that a road between certain specific points is necessary and proper; and leave the *particular* description of the road by routes and bounds, and by its course and distance, to the decision of the commissioners. 23 *Wend.* 360.

This certificate should be filed in the town clerk's office. It is a public document, open for the inspection of all the inhabitants of the town. 2 *Cow.* 623. It was decided by the Supreme Court, in this case also, that if such certificate came into the hands of a stranger, who is not a commissioner, the Supreme Court would compel him, by attachment, to file it in the town clerk's office, for inspection of a person who was a party to a suit where the road was in question.

Jurors not to be paid.

§ 74. No compensation shall be allowed any juror for examining and certifying in regard to the necessity and propriety of any highway being laid out, altered or discontinued, nor for appearing to make such examination. [*Sec.* 14 *of chap.* 180 *of* 1845.]

*Laws of 1834, chap. 267, § 1.

§ 75. [*Sec.* 62.] Before the commissioners shall determine to lay out the highway so applied for and certified, they shall cause notice in writing to be given to the occupant of the land through which the road is to run, of the time and place at which they will meet to decide on the application. The notice shall be served by delivering the same to such occupant, or, if he be absent, by leaving the same at his dwelling-house; and in either case, at least three days before the time of meeting. Notice to occupants.

Appendix, No. 38.

It would appear from this section, that a notice served on the *occupant* is sufficient. An occupant may have a very little interest in the premises. It is his duty, however, if a tenant, to notify his landlord. There are no decisions explaining this section.

§ 76. [*Sec.* 63.] The commissioners shall meet at the time specified in the notice, and shall hear any reason that may be offered for or against laying out the highway. If they shall determine to lay out such highway, they shall make out and subscribe a certificate of such determination, describing the road so laid out, particularly, by routes and bounds, and by its courses and distance, and shall deposit the same with the town clerk. Description of road.

Appendix No. 39.

Their determination must be confined to the highway applied for; but they are not limited to the precise *route*, specified in the application; they may in the exercise of a sound discretion, make such *variations* as they may judge proper. The departure must not, however, be so great as to induce the belief that the preliminary proceedings have been wholly disregarded; the general course of the road must be preserved. By the § 63 the applicant is to specify "*as near as may be, the route of the proposed highway.*" The freeholders act only upon the notice of the applicant, and the general description of the route therein, and have no authority to locate it with greater particularity. They determine only whether such highway is necessary and proper. The commissioners, by this section, are required to describe it *particularly, by routes and bounds, courses and distance. See Woolsey vs. Tompkins,* 23 *Wend.* 324. *Hallock vs. Woolsey,* (23 *Wend.* 328.)

The certificate required by this section will be, however, sufficiently conformable to it, if, in the description of the road, a *single line* is given which will be intended as the centre of the highway, and it contain a speeification of the quantity of land which the road will take for each proprietor over whose ground it passes. By this its width can be ascertained. *The people vs. Commissioners of Highways of Redhook,* 13 *Wend.* 310. *Herrick vs. Storer,* 5 *Wend.* 580.

It was also decided in *Woolsey vs. Tompkins,* 23 *Wend.* 324, to the same effect, that it will be sufficient if this certificate state the *termini* (this is its beginning and end) of the road, and its route by *courses and distance.* It is not necessary to state the bounds of each course.

It is advisable, however, that all the bounds and width, as well as its courses, be particularly stated to avoid future uncertainty or controversy.

§ 77. [*Sec.* 64.] The damages sustained by reason of the laying out and opening such road may be ascertained by the agreement of the owner and the commissioners of highways, provided such damages do not exceed one hundred dollars, and unless such agreement be made, or the owner of the land shall in writing release all claim to damages, the same shall be assessed in the manner prescribed by law, before such road shall be opened, or worked, or used. Every such agreement or release shall be filed in the town clerk's office, and shall forever preclude such owner from all further claim for such damages. [*Amended by* § 22 *of chap.* 455 *of* 1847.]

See Appendix, No. 40.

This agreement should be in writing, because it is required to be filed. The release must be in writing.

No one but the *owner* of the land is entitled to damages. The occupant or tenant has no claim. If he is disturbed in his possessions, he must look to his landlord for redress.

In the case of *The People vs. Supervisors of Oneida,* (19 *Wend.* 102.) the road was *wholly on the land of other persons,* but ran along the line of the applicant's farm, taking no portion of his land, but subjecting him to the expense of maintaining the whole of a *fence;* the expense of only *one-half* was before borne by him. The court decided that he was not entitled to damages. Justice Bronson says the case has not been provided for by Statute. The language of the Statute is broad enough to include consequential damages; but they can only be allowed where some portion of the applicant's land has been taken for the highway. The damages may, in certain cases, be ascertained by the agreement of the *owner* and the commissioners. If not agreed on, they are to be assessed in the manner pointed out in the following sections :

§ 78. Wherever any damages are now allowed to be assessed by law when any road or highway shall be laid out, altered or discontinued in whole or in part, such damages shall be assessed by not less than three commissioners to be appointed by the county court of the county in which such road or highway shall be, on the application of the commissioner or commissioners of the town; and the commissioners so appointed shall take the oath of office prescribed by the constitution, and shall proceed, on receiving at least six days notice of the time and place, to meet the highway commissioners and take a view of the premises, hear the parties and such witnesses as may be offered, before them; and they shall all meet and act and shall assess all damages which may be required to be assessed on the same highway, and shall be authorized to administer oaths to witnesses which may be produced before them under this section, and when they shall all have met and acted, the assessment agreed to by a majority of them, shall be valid; and when so made shall be delivered to a commissioner of highways of the town, who, within ten days after receiving it, shall file it in the town clerk's office. [*Amended by* § 2 *of chap.* 455, 1847.]

Damages, how to be assessed.

In considering the items of damage to be awarded to the owner, through whose lands the highway is laid out, the commissioners have a right to consider that the road will be devoted to the public as an *easement*, and also such right of pasturage by cattle, horses and sheep, as the electors of the town shall by their vote permit. *Griffin vs. Martin*, 7 *Barb.* 297.

It has been doubted by many whether the law (1 R. S. 341, § 5 Sub. 1.) empowering the electors of towns to make rules and regulations for ascertaining the sufficiency of fences, and for determining the times and manner in which "*cattle, horses and sheep shall be permitted to go at large on highways*" is constitutional. [*Ib. See Holladay vs. Marsh*, (3 *Wendall*, 142. *Tonawanda R. R. Co., vs. Munger*, (5 *Denio*, 255.) *White vs. Scott*, (4 *Barb. S. C. Rep.* 56.)

Appendix No. 36.

Provision in case persons conceive themselves aggrieved.

§ 79. Any person conceiving himself aggreived, or the commissioner or commissioners on the part of the town feeling dissatisfied by any such assessment, may, within twenty days after the filing thereof as aforesaid, signify the same by notice in writing, and serving the same on the town clerk and on the opposite party, that is, the persons for whom the assessments were made or the commissioner or commissioners of highways, as the case may be, asking for a jury to reassess the damages and specifying a time not less than ten nor more than twenty days from the time of filing said assessment, when such jury will be drawn at the clerk's office of an adjoining town of the same county by the town clerk thereof; which notice shall be served upon said opposite party within three days after service upon the town clerk as aforesaid, and may be served personally or by being left at the dwelling house of the party with some person in charge thereof, or if there be no such person, or the house be closed, then by fixing the same upon the outer door of said dwelling house. [*Sec.* 3, *of chap.* 455 *of* 1847.]

Appendix, No. 37.

§ 80. At the time and place mentioned in the preceding section, the town clerk of such adjoining town, having received three days previous notice that such jury is to be drawn, from the person or party asking a reassessment, shall deposit in a box the names of all such persons then residents of his town, whose names are on the last list filed in said town clerk's office of those selected and returned as jurors, pursuant to Article second, Title four, chapter seventh, part third of the Revised Statutes, who are not interested in the lands through which such road shall be located, nor of kin to either or any of the parties, and shall draw therefrom the names of twelve persons, and shall make a certificate of such names and the purposes for which they were drawn, and shall deliver the same to the party first asking for the reassessment. [*Sec.* 4 *of same chapter.*]

Names of jurors to be put in box and drawn.

Appendix No. 38.

§ 81. The party receiving such certificate shall, within twenty-four hours thereafter, deliver the same to a justice of the peace of the town wherein the damages are to be assessed; and it shall be the duty of such justice forthwith to issue a summons to one of the constables of his town, directing him to summon the persons named in said certificate, and shall specify a time and place in said summons at which the persons to be summoned shall meet, but no meeting of such persons shall be had within twenty days from the time of filing the assessment of damages in the town clerk's office by the commissioner or commissioners of highways. [*Sec.* 5 *of same chap.*]

Jury when to be summoned.

Appendix No. 39.

Twelve jurors to be drawn to reassess damages.

§ 82. Upon such persons appearing at the time and place mentioned in the summons, the justice who issued the summons shall draw by lot six of the persons attending to serve as a jury, and the first six persons drawn who shall be free from all legal exceptions, shall be the jury to re-assess all the damages required to be re-assessed upon the same highway; and the said jury shall be sworn by the said justice well and truly to determine and re-assess such damages as shall be submitted to their consideration, and shall take a view of the premises, hear the parties and such witnesses as may be offered by the parties, and sworn by said justice before them and shall render their verdict in writing under their hands, which shall be certified by said justice and be delivered to the commissioners of highways of the town, and the same shall be final. (*Sec.* 6 *of same chap.*)

Appendix No. 40, 41, 36.

Costs by whom to be paid.

§ 83. In all cases of assessments of damages under the provisions of this act by the commissioners appointed by a county court, the costs thereof shall be paid by the town in which the damages shall be assessed, and in cases of re-assessments of damages by a jury on the application of the commissioners of highways of any town, and the first assessment shall be reduced thereby, the costs of such assessment shall be paid by the party claiming the damages, otherwise by the town; and in case a re-assessment of damages shall be had on the application of the party for whom the damages were assessed, and such party shall fail to increase the same, he shall pay the costs thereof, but when such damages shall be increased by the jury the costs shall be paid by the town; and when applications shall be made by two or more persons for the re-assessment of damages by a jury, such jury shall be obtained in conformity with the terms of the notice first served upon the clerk of the town in which the damages are to be assessed; and all persons who may be liable for costs under this section shall be liable in proportion to the amount of damages respectively assessed to them by the first assessment, and may be recovered in an action of assumpsit at the suit of any person or persons entitled to the same before a justice of the peace. [§ 7 *of same chap.*]

Damages assessed to be audited by board of supervisors

§ 84. All damages which may be finally assessed or agreed upon by commissioners of highways for the laying out of any road except private roads, shall be laid before the board of supervisors by the supervisor of the town to be audited with the charges of the commissioners, justices, surveyors or other persons or officers employed in making the assessment and for whose services the town shall be liable, and the amount shall be levied and collected in the town in which the road is located, and the money so collected shall be paid to the commissioners of such town, who shall pay to the owner the sum assessed to him, and appropriate the residue to satisfy the charges aforesaid. [*Amended by* § 23 *of chap.* 455 *of* 1847.]

When such damages are finally assessed, the several persons in whose favor the same are awarded acquire a vested right to it, which it is not in the power of the town or commissioners to defeat by discontinuing the proceedings. *Hawkins vs Trustees of Rochester.* (1 *Wend.* 53.)

Damages in certain cases how estimated.

§ 87. [*Sec.* 71.] Where any person shall be the owner of any land over which any highway shall run, and such highway shall be discontinued, in whole or in part, by reason of some other road to be established and laid out under this Title, through the lands of the same person, the persons who shall assess the damages shall take into calculation the value of the road so discontinued, and the benefit resulting to such person by reason of such discontinuance, and shall deduct the same from the damages assessed for the opening and laying out of such new road; and thereupon the owner of the land may enclose so much of the highway so discontinued, as shall belong to him.

The fee of the land having always been in the owner of the soil, when the road is taken up it will revert to him, of course; and when the road is changed, in whole or in part, and a new location is substituted, it is right that the benefit should be deducted from the assessment of damages. Such deduction can be made only where the road or part of it is *discontinued by reason of the same, and the road established and laid* in substitution of the part discontinued. *Jackson vs. Hathaway*, (15 *Johns. R.* 447. *Whittock vs. Cook*, 15 *ib.* 491. *See also notes under* [*Sec.* 126 *and* 127.] *Also. Cortelyon vs. Van Brundt*, 2 *J. R.* 357.)

The public use of a highway being but an easement, subject to which the owner of the land over which it passes retains his title, there is always a contingency by which the owner may return into full possession of the land, on its being no longer required by the public. When this contingent event will happen is ordinarily unknown, and is wholly immaterial, as regards the rights of the land holder. Whether the public retains the use of the land for a century, or for a year, or but for a single day, cannot effect his title to a compensation. That becomes fixed and vested, the instant his property is taken for public use. *The People vs. Supervisors of Westchester.* (4 Bar. 64.)

Disagreements respecting certain roads.

§ 88. [72.] When the commissioners of highways of any town shall disagree with the commissioners of any other town in the same county, relating to the laying out of a new road, or the alteration of an old road, extending into both towns; or when the commissioners of a town in one county shall disagree with the commissioners of a town in another county, relative to laying out a new road, or altering an old road which shall extend into both counties; the commissioners of both towns shall meet together at the request of either disagreeing commissioners, and make their determination upon such subject of disagreement.

§ 89. [73.] Whenever it shall become necessary to have a highway upon the line between two towns, such highway shall be laid out by two or more of the commissioners of highways of each of said towns, either upon such line, or as near thereto as the convenience of the ground will admit; and they may so vary the same either to the one or the other side of such line, as they may think proper. Road upon line of two towns.

§ 90. [74.] It shall be the duty of the same commissioners, when they lay out such highway, to divide it into two or more road districts in such manner that labor and expense of opening, working, and keeping in repair such highway, through each of the said districts, may be equal, as near as may be, and to allot an equal number of the said districts to each of the said towns. How divided into districts.

§ 91. [75.] Each district shall be considered as wholly belonging to the town to which it shall be allotted, for the purpose of opening and improving the road, and for keeping it in repair; and the commissioners shall cause such highway, and the partition and allotment thereof, to be recorded in the office of town clerk in each of their respective towns. Effect of allotment.

§ 92. [76.] All highways heretofore laid upon the line between any two towns, shall be divided, allotted, recorded, and kept in repair, in the manner above directed. Former road.

PRIVATE ROADS.

Private roads how laid out.

§ 93. [77.] Private roads may be laid out on application being made to the commissioner or commissioners of highways of any town, and they or he shall proceed in the manner provided in the tenth section of an act passed December 14, 1847, entitled "An act to amend an act entitled an act to reduce the number of town officers, and town and county expenses, and to prevent abuses in auditing town and county accounts, passed May 10, 1845," to give notice and summon a jury, and all the proceedings required by said tenth section shall be had, and the jury shall determine the necessity of the road, and the amount of damage to be sustained by the opening thereof, and such amount together with the expenses of the proceedings, shall be paid by the person to be benefited. [*Sec.* 1 *of chap.* 77 *of* 1848.]

This provision having been pronounced unconstitutional by the Supreme Court, the new constitution has provided as follows:—

"*Private roads may be opened in the manner to be prescribed by "law, but in every case the necessity of the road and the amount of all "damage to be sustained by the opening thereof, shall be first deter-"mined by a jury of freeholders, and such amount together with the "expenses of the proceeding shall be paid by the person to be benefit-"ted.*" Art. 1, Sec. 7.

This provision was introduced on account of the decision (4 *Hill* 140 *Taylor vs. Porter & Ford.*) in which the Supreme Court decided that the laws in relation to private roads were contrary to the former constitution.

The Supreme Court had declared that a private road might be laid under this section by the *consent* of the owner of the land, on the principle that a man may renounce a constitutional provision made for his own benefit. *Baker vs. Braman*, (6 *Hill*, 47.)

Provision in case of damages for laying out private roads.

§ 94. In all cases of assessments of damages for laying out or altering any private road, the commissioners of highways of the town where such road may be situated, shall serve a notice in the same manner as prescribed in § seventy-nine of this act, upon the town clerk of the town, and upon the persons interested in said road, specifying a time when a jury of the town will be drawn by the said town clerk at his office, which time shall be within not less than six nor more than ten days from the time of service of such notice. At the time mentioned in such notice, said clerk shall proceed to draw in the same manner from the jury list last filed, the same number of names as provided for juries in cases of reassessments, under section eighty of this act; and the names so drawn shall be certified to, and the jurors summoned, drawn and sworn, and all proceedings had, and the verdict shall be made, certified to, and be filed as is provided in cases of reassessments of damages as aforesaid; and the same jury shall assess all damages required to be assessed for the same road. (*Sec.* 10 *of chapter* 455 *of* 1847.)

The notice should be in writing. If the party interested should appear without notice, it would waive all irregularity, and he could not afterwards object, according to the decision in *The Mohawk R. R.* vs. *Artcher*, 6 *Paige*, 83.

In the same case the chancellor decided that the damages should be actually paid before the property is taken.

"It is evident that it never was the intention of the legislature, to authorize the opening of a private road, through the lands of one person for the benefit of another, until after the damages had not only been assessed, but actually paid to the person, for whose benefit such private road is laid out and opened." 6 *Paige*, 83.

See Appendix, Nos. 45, 46.

§ 95. [*Sec.* 79.] Every such private road, when so laid out, shall be for the use of such applicant, his heirs and assigns; but not to be converted to any other use or purpose, than that of a road. Nor shall the occupant or owner of the land through which such road shall be laid, be permitted to use the same

For what purpose road to be used.

14 John. 383.

5 Wend. 580.

as a road, unless he shall have signified his intention of so making use of the same to the jury or commissioners who ascertained the damages sustained by laying out such road, and before such damages were so ascertained.

The applicant has the sole and exclusive right to use the road; unless the owner at the time of laying out of the road signified his intention to make use of it. The applicant may maintain tresspass against the owner or any other person for using the road. *Lambert vs. Hoke*, 14 *Johns.* 383.

Where a party obtains a right to a *private road* of the width of two rods, the owner of the land through which it passes, must so build his fence, as to leave full two rods in width in every part of the road; he cannot build a *Virginia fence*, placing the center of the exterior lines of the two rods with the angles projecting into the road. *Herrick vs. Storer*, 5 *Wend.* 580.

A party will be deemed to have consented to such location of the fences, if apprised that the damages of the owner of the lands were assessed in reference to such location, or if he permits the fences to be thus built without objection. *Herrick vs. Storer*, 5 *Wend.* 580.

Width of roads.

§ 96. [*Sec.* 80.] All public roads to be laid out by the commissioners of highways of any town, shall not be less than three rods wide, and all private roads shall not be more than three rods wide.

The record of a private road laid out by the commissioners, designating the course, distance and quantity of land taken, is sufficiently definite to show that the road was intended to be two rods wide; and parol evidence that such would be the result from the data given, is admissible. *Herrick vs. Storer*, 5 *Wend.* 580.

Where roads are claimed because they have been used twenty years, they may be more or less than three rods wide; that will depend on their actual *user*. When public roads are laid out under the Statute, they will be deemed to be at least three rods in width, and a person will be liable for an obstruction within that width. *Harlow vs. Humuston*, 6 *Cow.* 189.

Old roads how discontinued.

§ 97. [*Sec.* 81.] Whenever application shall be made for the discontinuance of an old road, on the ground that it has become useless and unnecessary, the commissioners of highways to whom such application shall be made, shall summon twelve disinterested freeholders of the town, to meet on a day certain, to consider such application. Such freeholders when met, shall be sworn well and truly to examine and certify in regard to the propriety of such discontinuance.

See Appendix, Nos. 49, 50.

Any person liable to perform highway labor may make the application. *See* § 58 *and Notes.*

The commissioners need not issue any precept to summon the jury. They are bound to select the jury themselves, and cannot delegate that authority to another. *The People vs. Commissioners of Greenbush,* 24 *Wend.* 367.

§ 98. [*Sec.* 82.] They shall then proceed to view such road, and if they shall be of opinion that the same is useless and unnecessary, they shall make and subscribe a certificate in writing to that effect, which shall be delivered to the commissioners of highways, who shall thereupon proceed to decide upon such application. Ibid.

See Appendix, No. 51.

The commissioners have no power in any case to discontinue a road without the oath of the freeholders. After the freeholders have made the certificate, the commissioners may grant the application or refuse it. An appeal to the judges may be had from their decision. For form for decision of commissioners, *See Appendix, No.* 52.

§ 99. [*Sec.* 83.] Applications, certificates and other papers relating to the laying out, altering, or discontinuing of any road, shall be filed by the commissioners of highways, as soon as they shall have decided thereon, in the office of the town clerk of the town. Papers where filed

This section should be carefully observed, as it may save much expense and litigation.

APPEALS.

Right of appeal.

§ 100. [*Sec.* 84.] Any person who shall conceive himself aggrieved by any determination of the commissioners of highways, either in laying out, altering or discontinuing any road, or in refusing to lay out, alter or discontinue any road, may at any time within sixty days after such determination shall have been filed in the office of the town clerk, appeal to the county judge of the county in the same manner as appeals were heretofore allowed to be brought to three judges under Title first, Article fourth, chapter sixteenth, Part first of the Revised Statutes; and when any appeal shall be brought under this section, the said judge, or in case of his residence in the town, or of his interest in the lands through which the road shall be laid out, or in case he is of kin to any of the persons interested in said lands, or in case of his disability for any cause, then one of the justices of the sessions shall, after the expiration of the said sixty days, appoint in writing three disinterested freeholders who shall not have been named by the parties interested in the appeal, and who shall be residents of the county, but not of the town wherein the road shall be located, as referees to hear and determine all the appeals that may have been brought within the said sixty days, and shall notify them of their appointment, and deliver to them all papers pertaining to the matters referred to them. Upon receiving notice of an appointment the said referees shall possess all the powers and discharge all the duties heretofore possessed and discharged by the three judges, and give the same notices heretofore required to be given under title first, article four, chapter six, part one aforesaid, and before proceeding to hear the appeal or appeals, they shall be sworn by some officer authorized to take affidavits to be read in courts of record, faithfully to hear and determine the matters referred to them. [*Sec.* 8, *of chap.* 455 *of* 1847.]

This section provides that the referees "*shall possess all the powers and discharge all the duties heretofore possessed and discharged by the three judges.*

The former section was in these words :

"Where an appeal shall have been made from a determination of commissioners refusing to lay out or alter a road, and the judges shall reverse such determination, such judges shall lay out or alter the road applied for; and in doing so, shall proceed in the same manner in which the commissioners of highways are directed to proceed, in the like cases. Such road shall be opened by the commissioners of the town, in the same manner as if laid out by themselves."

Under this section there were several decisions. In cases where freeholders have made a certificate, and commissioners have refused to lay the road, the judges cannot affirm *in part* and reverse *in part.* 25 *Wend.* 453.

The power of the judges is strictly appellate, and they cannot lay out a road different from the one applied for. It must be substantially the same road designated in the application. *Ex parte, commissioners of Danube,* 1 *Cow.* 142.

The judgment of three distinct bodies of men can be had, if desired, by the party interested, before the proposed road can be established: 1. The *freeholders;* 2. the commissioners; and 3. the judges: and two of these bodies must concur in the laying out of the highway. The certificate of the freeholders is indispensable, and the commissioners are confined to the *termini* of the route as certified by the freeholders. Indeed, if it be not so, the preliminary proceedings of the freeholders are of little importance; as the fact that they have recommended a given route as necessary for the public convenience, affords no ground for the presumption that they would have certified to a part of it; the same remark is equally applicable to the judges. To give any efficient force to the action of the freeholders, the power of the commissioners and judges in the matter must be limited to the line as recommended. *Coms. of Sherburne vs. Judges of Chenango,* (25 *Wend.* 453.)

In *the laying out of a road,* after the refusal of commissioner, the judges are authorized to reject all *parol evidence* as to the route of the road designated by the *freeholders* and passed upon by the *commissioners,* and may limit themselves to the original *application* for the road, the *certificate* of the freeholders, and the *adjudication* of the commissioners. *The People vs. Judges of Dutchess,* (23 *Wend* 360.)

If an error happen in the description of the road laid out, the officers laying out the same may be required, by *mandamus*, to file a new certificate, and give a *description* conformable to the facts of the case. *Woolsey vs. Tompkins*, 23 *Wend.* 324.

Where errors have occurred in the description of a road, in the order of the judges reversing the determination of commissioners in laying out a road, it is competent for the judges, after the filing of their order, to deposit in the town clerk's office a document correcting the errors; and such document will be considered as an amendment of the order; but the judges are not permitted to review and alter what they have done *judicially*. The making up the *certificate* is a mere *ministerial* act. 23 *Wend.* 324.

In the case of *Lawton vs. Coms. of Cambridge*, (2 *Caines*, 179,) it was desided that the authority of the judges to hear the appeal was confined to the *merits* alone—the necessity and propriety of laying the road, or refusing so to do. They reiterated and confirmed this decision in 13 *Wend.* 432. *Coms. of Warwick vs. Judges of Orange.*

Justice Nelson says: I perceive no solid reasons against the position laid down in the case of *Lawton vs. Coms. of Cambridge*, that the authority of the judges to hear the appeal is confined to the merits alone; on the contrary, much inconsistency and violation of several principles is avoided, and the spirit of the Statute faithfully executed, by confining them to that question. The power of this court can be legitimately exerted to keep both the commissioners and the judges within their proper bounds in conducting their proceedings; and then the determination, first by the one, and then by the other, on appeal, will be conclusive in the premises, according to the letter of the Statute. Upon any other view, a road might be laid out, upon the reversal of the determination of the commissioners, on mere legal questions, without involving their opinion upon the merits. The act contemplates such an opinion before the judges can act.

Certiorari.—Either party, if dissatisfied with the decision of the judges, may bring a *certiorari* in the Supreme Court to review their determination; to which the judges will then return as in 1 *Cow.* 23. The party makes an affidavit of the facts and the error on which he relies, and applies to the court at special term. It is called a *common law certiorari.* But on the return of the judges, no other question can be raised than those relating to the *jurisdiction* of the judges, and the regularity of their proceedings. *Birdsall vs. Philips*, (17 *Wend.* 464.) *Prindle vs. Anderson*, (19 *Wend.* 391.)

"The supervisory power of the Supreme Court over inferior tribunals by means of this writ, except in cases where special provisions have been made by the legislature, only extends to questions touching the jurisdiction of the subordinate tribunal, and the regularity of its proceedings. If they neither exceed their powers, nor depart from the forms prescribed to them by law, their decision upon the merits of any controversy before them is final and conclusive." *The People vs. Judges of Dutchess*, (23 *Wend.* 360.)

The appeal must be made and served within sixty days after the filing of the order.

An appeal lies only in the cases enumerated in this section; no appeal can be had from a decision under § 103. The right of appeal is not given in general terms, but the cases in which it may be exercised are specially enumerated. There is a reason for not making the right of appeal coextensive with the powers of the commissioners. In opening new roads, and altering and shutting up old ones, the commissioners act judiciously, and in matters materially affecting the interests of individuals. But in describing and recording a road that has already become public by twenty years' use, they perform little more than a ministerial duty. *The People vs. the Judges of Cortland*, 24 *Wend.* 491.

An appeal cannot be made to judge who was one of the freeholders and signed the certificate, 2 *R. S. p.* 275, § 3. *Commissioners of Carmel vs. Judges of Putnam*, 7 *Wend.* 264. If this objection was not urged on the hearing of the appeal, it is not such an error as will produce a reversal of the judges' decision on *certiorari.* 7 *Wend.* 264.

An appeal may be waived by the applicant. 1. By giving notice to that effect, and thereby withdrawing his appeal; or 2. By his conduct: as in the case of *Lansing vs. Caswell*, 4 *Paige*, 519, when after an appeal, the judges not proceeding, the commissioners went on and empannelled a jury to assess the damages, and the appellant appeared before the jury and litigated the question as to the amount of damages, and subsequently applied to the board of supervisors to increase the amount of damages allowed by the jury; it was held that the appeal was thereby waived.

As many different persons as are affected by the decision may appeal, and if they choose, all may bring separate appeals.

Form of appeal.

§ 102. [*Sec.* 86.] Every such appeal shall be in writing, addressed to the judge, and signed by the party appealing. It shall briefly state the ground upon which it is made, and whether it is brought to reverse entirely the determination of the commissioners, or only to reverse a part thereof, and in the latter case it shall specify what part. (*Amended by* § 8, *chapter* 455 *of* 1847.)

See form, Appendix, No. 53.

On an appeal from the decision of the commissioners in *refusing to lay out* an highway, the judges cannot *reverse* in part, and *affirm* in part, except in those cases where the certificate of the freeholders is not requisite. That part of the above section specifying the requisites of an appeal, as to whether it is brought to reverese *in whole* or *in part*, applies only to cases where the roads are laid out without the intervention of freeholders. *Commissioners of Sherburne vs. Judges of Chenango*, 25 *Wend.* 453.

Proceedings.

§ 103. [*Sec.* 87.] It shall be the duty of the referees, to proceed thereon as soon as may be convenient. Where the determination appealed from was against an application for laying out, altering or discontinuing a road, the referees shall give notice to the commissioners by whom such determination was made. When the appeal is from a determination in favor of an application for laying out, altering or discontinuing a road, the notice shall be given to the commissioners, and to one or more of the applicants for such road. In all cases, the notice shall specify the time and place at which the referees will convene to hear the appeal. (*Amended by* § 8 *of chapter* 455 *of* 1847.)

See Appendix, No. 54.

If the judges do not, or will not proceed, they may be compelled by *mandamus* to hear and decide the appeal. *Lansing vs. Caswell*, 4 *Paige*, 519.

Without this notice the referees have no jurisdiction to proceed on the appeal. *The People vs. Judges of Herkimer*, 20 *Wend.* 186.

This case was an appeal to the judges under the former statute, and the court decided "that the judges had no jurisdiction to proceed on the appeal, until a notice of eight days, of the time and place of meeting, had been served on the commissioners.

Notice of appeal.

§ 104. [*Sec.* 88.] Every such notice shall be served at least eight days before the time mentioned therein, by delivering the same to one of the commissioners whose determination is appealed from, or by leaving the same at his dwelling-house. If the notice be also directed to an applicant, it shall be served in the same manner.

This notice must be in writing. *Gittrel* vs. *Columbia, Tompkins Co.* 1 *Johns. Cas.* 107, *and* 20 *Wend.* 186.

Notice need only to be delivered to one of the commissioners, it is notice to all; and the one who receives it should give information to the other commissioners. *The People* vs. *Judges of Herkimer*, 20 *Wend.* 186.

If all the commissioners should meet with the judges, it would probably waive the omission of written notice; but without such notice the appearance of one commissioner would not be sufficient.—*Ibid.*

§ 105. [*Sec.* 89.] It shall be the duty of the referees to convene at the time and place mentioned in the notice, and hear the proofs and allegations of the parties. They shall have power to issue process to compel the attendance of witnesses, and may adjourn from time to time, as may be necessary. Their decision, or that of any two of them, shall be conclusive in the premises; and every such decision shall be reduced to writing, signed by the referees making it, and be filed by them in the office of the town clerk of the town, who shall record the same.—(*Amended by* § 8 *of chapter* 455 *of* 1847.) Proceedings.

Under the former law, when the decision was made by the judges, it was decided that the three judges should meet and deliberate; or if all are notified, and one refuses or neglects to appear, the others may propound, deliberate, and decide. *Woolsey vs. Tompkins*, 23 *Wend.*, 324.

The same principle will hold as to referees. In the last case the court said, "it abundantly appears from the return, that the three judges met and deliberated upon the proceedings had under the appeal, and the decision of the two is therefore valid and binding. It has been decided by the court, as a general principle, that where an authority is conferred on three, a majority may act, although the matter under a determination does not recite that the three were present and participated in the proceeding, or had notice and refused, the court will presume the fact till the contrary appear." *Ib.*, 17 *Johns.* 461; 9 *Wend.* 17.

On the hearing, an inquiry into the damages of the owners of the land is proper, to enable the referees to determine, whether the benefit will equal the expense, and whether the public good will be promoted by the road. *Commissioners of Bushwick vs. Mesreole*, 10 *Wend.* 122.

A commissioner is not a competent witness on such hearing.—*Ibid.*

A common law certiorari will undoubtedly lie, from the decision of the referees, on good causes shown to the supreme court. *Kinderhook vs. Claw*, 15 *Johns. R.* 537; *Lawton vs. Commissioners of Cambridge*, 2 *Cain's Rep.* 179; *Clark vs. Phelps*, 4 *Cowen, Rep.* 190.

For form for return to certiorari, *see* 1 *Cowen Rep.* 23. *The People vs. Commissioners of Salem. See Appendix, No.* 53.

Fences to be remov'd

§ 111. [*Sec.* 96.] Whenever the commissioners of highways shall have laid out any public highway, through any enclosed, cultivated or improved lands, in conformity to the provisions of this Title, and their determination shall not have been appealed from, they shall give the owner or occupant of the land through which such road shall have been laid, sixty days' notice, in writing, to remove his fences. If such owner shall not remove his fences within the sixty days, the commissioners shall cause such fences to be removed, and shall direct the road to be opened and worked.

Fences to be remov'd

§ 112. [*Sec.* 97.] If the determination of the commissioners shall have been appealed from, then the sixty days' notice shall be given, after the decision of the referees upon such appeal shall have been filed in the office of the town clerk of the town.—(*Amended by* § 8 *of chap.* 455 *of* 1847.)

Sixty days' notice must be given before proceeding to open the road, as well when it has been established by an alteration made by judges, after the same has been laid by them on appeal, as when a road is originally laid out by commissioners. Actual notice must be shown in such cases, as it will not be presumed. *Case* vs. *Thompson* 6 *Wend.* 634.

An appeal from the decision of the commissioners, laying out a road through "*enclosed, cultivated or improved lands,*" is in effect, a stay of proceedings; and if the commissioners go on and open the road, pending an appeal, they will be tresspassers. But when they lay out a road through unimproved lands, or when they alter or discontinue an old road, there appears to be no provision for a stay of proceedings. *Drake* vs. *Rogers*, 3 *Hill*, 604.

If, on an order being made discontinuing a road, a fence be built across it, an appeal, subsequently brought, will not have the effect of rendering the fence a *public nuisance*. 3 *Hill*, 604.

Final determination to be carried in to effect by the commissioners.

§ 116. Whenever there shall have been a final determination upon any appeal or appeals provided for as aforesaid, making it necessary that any road or highway shall be laid out, altered, opened or discontinued, it shall be the duty of the commissioner or commissioners of highways of the town where the same is to be done, to carry out such determination the same as if the decision of such commissioner or commissioners had been in favor thereof, and there had been no appeal. (*Sec.* 13, *chap.* 180 *of* 1845.)

Pay of referees.

§ 117. Every referee appointed under the preceding section shall be entitled to receive two dollars for every day employed in the hearing and decision of such appeal or appeals, to be paid by the party appealing, where the determination of the commissioners shall be confirmed, but where it is reversed, to be a charge upon the county; and when said referees shall make any decision, laying out, altering, or discontinuing any road in whole or in part, it shall be the duty of the commissioners of highways of the town to carry out such decision in the same manner as required in cases of final determinations of appeals as provided by the thirteenth section of the act hereby amended, and such decision shall remain unaltered for the term of four years from the time the same shall have been filed in the office of the town clerk. (*Sec* 9, *of chap* 455, *of* 1847.)

Certain acts of commissioners confirmed.

§ 118. [*Sec.* 98.] The acts and doings of the commissioners of highways of the several towns in this state, or of any two of them, in laying out, altering, or discontinuing any road or highway, since the thirty-first day of December, one thousand eight hundred and five, and prior to the fourteenth day of April, one thousand eight hundred and twenty-six, are confirmed from the last mentioned day; provided such commissioners, or any two of them, shall have caused a survey of such roads or highways to be filed or recorded in the office of the town clerk of the town. But such conformation shall not affect any decision of the judges of common pleas, made prior to the fourteenth day of April, one thousand eight hundred and twenty-six, confirming or reversing the determination of the said commissioners; nor any appeal from such determination, made within six months after that day; nor any suits or proceedings which on that day were pending, at law or in equity.

This section is designed to confirm the acts of the commissioners previous to the time therein mentioned, although there may have been an *irregularity* in their proceedings.

So where the commissioners, in 1821, caused a survey of the road to be filed and recorded, the existence of the road as a public highway was adjudged by the court to be fully established, although it did not appear that the owner had given his consent, or that a certificate had been made by freeholders; it being an irregularity that was cured by this section. *Parker* vs. *Van Houton*, (7 *Wend.*) *Laws of* 1826, *p.* 229, § 8.

When roads cease.

§ 119. [*Sec.* 99.] Every public highway already laid out, that shall not have been opened and worked within six years from the time of its having been so laid out, and every such highway hereafter to be laid out that shall not be opened and worked within the like period, shall cease to be a road for any purpose whatever.

The road must be both opened *and worked.* It is not necessary that every part of it should be worked. This might be impossible.

Where the country is new, and the population thin, there might not be highway labor sufficient in the district to work all the road.

It should be embraced in a road district, and worked as far as there is necessity for work or highway labor to expend upon it.

This section differs from the act of 1813. *See Lyon* vs. *Munson*, 2 Cowen, 426.

§ 120. [*Sec.* 100.] All public highways, now in use, heretofore laid out and allowed by any law of this state of which record shall have been made in the office of the clerk of the county or town; and all roads not recorded, which have been or shall have been used as public highways for twenty years or more, shall be deemed public highways, but may be altered in conformity to the provisions of this Title. What roads highways.

This enacts what the common law has already declared, that roads, although not recorded, which have been used as public highways for twenty years, shall be deemed public highways. *The People* vs. *Judges of Cortland* 24 *Wend.* 490. *See the People* vs. *Lawson*, 17 *Johns. R.* 276. *Colden*, vs. *Thurber*, 2 *Johns R.* 424.

In 7 *Johns.* 106, *Galatin* vs. *Gardiner*, it was decided that when a road had been used twenty years, leading to a dock and landing, or ferry, and such ferry had been changed, and some part of the way had been appropriated and built upon, the passage still continuing open to the dock and landing, such road still remained a public highway.

§ 121. [*Sec.* 101.] It shall be the duty of the commissioners of highways, to order overseers of highways to open all roads to the width of two rods at least, which they shall judge to have been used as public highways for twenty years. Width.

This does not authorize the commissioners to adjudge what was *originally intended*, in relation to the width or location of the road, any further than such intention has been manifested by permitting the way to be used. It is a power in relation to the road *as it actually exists and has existed* for the last twenty years. It does not authorize the commissioners to create or enlarge, but only to pepetuate the evidence of public right. Both the extent and the fact of dedication depend on the user. *The People* vs. *the Judges of Cortland.* 24 *Wend.* 491.

There is no *appeal* from the decision of the commissioners acting under this section. So when the commissioners of Cortlandville enlarged a road describing it as *three rods* wide, when in fact its width never exceeded *two rods*, and also changed its location, it was held that the aggrieved had no remedy by *appeal* the judges of the county in which the road is situated. Same in 24 *Wend.* 491.

There is a doubt expressed what remdey the party injured would have; whether a *common law certiorari* in the supreme court, or whether the proceedings of the commissioners should be treated as *void*, and all persons concerned treated as tresspassers.

SPECIAL COMMISSIONERS.

The legislature in the act of April 18, 1838, among other things, conferred the power upon the board of supervisors in each county, at their annual meeting and when lawfully convened—

4. "To appoint special commissioners to lay out public highways in those cases where they shall be satisfied that the road applied for is important, and that the authority now conferred by law upon commissioners of highways, cannot or will not be exercised to accomplish the laying out of such road."

AN ACT *to amend an act entitled "An Act to enlarge the powers of Boards of Supervisors," passed April* 18, 1838.

Passed April 3, 1848.

The People of the State of New York, represented in Senate and Assembly, do enact as follows:—

§ 1. The power given to boards of Supervisors by subdivision four, of section one, of the act entitled "An Act to enlarge the powers of boards of Supervisors" passed April 18, 1838, to appoint special commissioners to lay out public highways, shall not be exercised by any board of supervisors, unless the applicant therefor shall prove to such board of supervisors the service of a notice in writing, on a commissioner of highways of each town, through and into which any such highway is intended to be laid, at least six days previous to presenting such application specifying therein the object thereof, and names of persons proposed to be appointed such commissioners.

§ 122. Whenever any turnpike corporation shall become dissolved, or the road discontinued, its road shall become a public highway, and be subject to all the legal provisions regulating highways. [*Sec.* 1 *of chap.* 262 *of* 1838.] When turnpike roads to become public highways.

ARTICLE FIFTH.

Regulations and Penalties concerning the Obstructions of Highways, and Encroachments thereon.

SEC. 123. Five dollars forfeiture for obstructing highway.
124. Where fences encroach on highways, commissioners to order them removed.
125. If not removed as required, party to forfeit fifty cents a day until removed.
126. If encroachment denied, jury to be summoned to inquire into premises.
127. Jury to be sworn, and to hear proofs and allegations.
128 & 129. Finding of Jury, how enforced; who to pay costs, &c.
130. No fences to be removed but between the first of April and November.
131 & 132. Side walks may be made along road, penalty for injuring, &c.
133. When trees fall into highway, occupant to remove them; penalty for neglect.
134. Persons cutting trees without consent of occupant, to forfeit one dollar, &c.
135. Five dollars penalty for leaving trees in any waters declared a highway.
136. No swinging gates allowed, except on lands liable to be overflowed.
137. Such gates to be maintained by the person benefited.
138. If more than one, expense borne by all the occupants benefited thereby.
139. Overseer of district in which such gates are, to file statement of charges, &c.
140. Overseers to collect such charges from persons bound to pay them.
141. Commissioners to file an account of such gates in town clerk's office; penalties.

§ 123. [*Sec.* 102.] Whoever shall obstruct any highway, or shall fill up or place any obstruction in any ditch coustructed for draining the water from any highway, shall forfeit for every such offence the sum of five dollars. Penalty for obstructing.

OBSTRUCTION.—Any thing that impairs or impedes the exercise of the public right to travel upon a public highway is a nuisance. 1 *Hawk. P. C. Book* 1, 76, § 48.

These are principally laying stone, wood, timber, or other things in or across the highway; by digging ditches, or otherwise injuring the surface of the road; by building fences, walls, or building into or across the highway, and a great variety of ways which it is unnecessary to repeat.

It is an interruption for any person to occupy a street for an unreasonable time, although he be in the exercise of his ordinary calling. *Woolrych on Ways*, 47.

Any obstruction of a highway for an unreasonable length of time although in the prosecution of a lawful business, as in the loading or unloading wagons or drays, or by stages waiting and soliciting for passengers, is indictable as a public nuisance, although room enough might still be left for the accommodation of the public on the opposite side of the street. The public are entitled to the use and enjoyment of the whole of the highway, and no individual can appropriate a portion of it to his own exclusive use, and shield himself from responsibility by saying that enough is still left for the accommodation of others. *Hart vs. Mayor &c. of Albany*, (9 *Wend.* 571.) 6 *East.* 427, 3 *Camp.* 224, *Ibid.* 229.

There are three ways for removing obstructions or punishing the offender: 1. By indictment; 2. Sueing for penalty; and 3. Abating it as a nuisance; and in addition to these remedies the party injured may have an action on the case against the person who interrupted the highway.

Indictment.—Any unauthorized obstruction of a highway to the annoyance of the public, is an indictable offence. *Russell on Crimes*, 463, 3 *Camp.* 227. *Wetmore vs. Tracy*, 14 *Wend.* 250, *and cases there referred to.* All injuries to the highway as by digging a ditch or making a fence across it; or laying logs of timber in it; or any other acts which will render it less commodious to the citizens, are nuisances, at common law. 1 *Hawk. P. C. c.* 76, § 144, *Russell*, 461.

Penalty.—Debt is the proper form of action for penalty. *Cowen's Treatise, p.* 42. The commissioners must sue in their names, adding their title. *Commissioners of Cortlandville vs. Peck*, 5 *Hill.* 215.

A plea of title is no bar to an action under this section. *Parker vs. Van Houton*, 7 *Wend.* 145, also *Fleet vs. Young*, 7 *Wend.* 291.

While the law gives redress for any injury to a private way, an action for obstructing one which is public will only lie, where special damages have been sustained. *Lansing vs. Wiswall*, 5 *Denio* 213, 1 *Ch. Pl. e d* 1837. *Woolr. on Ways*, 52—54. *Lansing vs. Smith*, 8 *Cow.* 156.

In this suit special damages must be *alledged and proved.* 5 Denio, 52.

An individual may sue for an obstruction to a public highway when special damages have resulted to him; as where he has expended money in removing obstructions so that he could travel on the road. *Ibid.*

Where a plaintiff proved that a road leading from his dwelling house across the lands of others to a village, had been used for travelling for more than forty years, and that the defendant, who occupied lands through which it ran had recently obstructed it, *held* that the plaintiff was entitled to recover, though it appeared that it had formerly been part of a public highway, and the commissioner of highways, more than twenty years before the commencement of the suit, had changed the route of the highway so as to discontinue it at the place where the obstructions were put, it also appearing that the plaintiff and his predecessors in the title to the farm which he owned had continued to use and enjoy it notwithstanding such discontinuance.

Twenty years' uninterrupted and unqualified enjoyment of a way across the lands of another, is decisive evidence of a grant of the right of way. *Lansing vs. Wiswall*, 5 *Denio*, 213.

Under the provision of the statute making it the duty of the justice to dismiss the cause where it appears on the trial, from the plaintiff's own showing, that the title to lands is in question, it has been held in *Willoughby vs. Jenks* 20 *Wend.* 96, that an action of trespass for digging up the soil in the street, brought by the owner of a lot bounding on the street, which had been dedicated to the public, but which had not been accepted or recognized by the local officers as a public street, presented a question of title to land, and could not be tried in a justice's court. In that case, the street not having been used by the public for twenty years, it was necessary for the plaintiff, in order to show title in himself, to prove certain explicit acts of dedication, a map laying down the street, an agreement among the proprietors that it should be a public street, and his own title deeds bounding him upon it as such; but this evidence the Supreme Court, on error, held that the justice had no right to consider under the provision of the statute above referred to. In that case the court further held that proof of actual possession of a lot adjoining a street, shows a constructive title in the occupant to the soil extending to the centre of the street; but to give such title, the street must have been accepted by the public as such; although dedicated, if not accepted, it remains the property of the original proprietor, subject to the easement or right of way of the purchasers of lots adjoining the street.

Fences when and how to be removed.

§ 124. [103.] In every case where a highway shall have been laid out, and the same has been or shall be encroached upon by fences, erected by any occupant of the land through or by which such highway runs, the commissioners of highways of the town, shall, if in their opinion it be deemed necessary, order such fences to be removed, so that such highway may be of the breadth originally intended. The commissioners making the order, shall cause the same to be reduced to writing and signed. They shall also give notice in writing to the occupant of the land, to remove such fences within sixty days. Every such order and notice shall specify the breadth of the road originally intended, the extent of the encroachment, and the place or places in which the same shall be.

Appendix No. 60.

The extent of the encroachment must be given with precision in the notice and order. *In Mott vs. Commissioners of Rush,* (2 *Hill,* 472,) it was held that an order and notice describing the encroachments of "the average width of one rod or upwards," was insufficient. "The statute," says the court, "is very explicit, and for a very obvious reason, viz—to enable the party, if he see fit, to comply with the order at once." The description should be full and precise, so as to fix the place and extent of the encroachment beyond all doubt or embarrassment to the occupant, for a heavy penalty is annexed in case of refusal to comply. A notice directing the removal of the fence, "*so that the highway might be of the width originally intended,*" is not good under this section.

In order to perform their duties under this section, the commissioners must meet and deliberate together on the subject of the alleged encroachment, and a majority may make the order. *Spicer vs. Slade,* (9 *John. R.* 359.)

Penalty. § 125. [*Sec.* 104.] If such removal shall not be made within sixty days after the service of such notice, the occupant to whom the notice shall be given, shall forfeit the sum of fifty cents for every day, after the expiration of that time, for which such fences shall continue unremoved. Expense of removal. And the commissioners of highways may remove or cause to be removed such encroachment, and the occupant of the premises shall pay to the commissioners of highways all reasonable charges therefor, to be collected in the manner provided in the forty-fifth section of said title.—[*Amended by* § 1 *of chap.* 300 *of* 1845.]

This section, authorizing commissioners to order the removal of fences, by the erection of which highways have been encroached upon, does not abrogate the common law remedy of abatement of nuisance by the mere act of individuals, or abolish the remedy by indictment; the remedy given by this section is cumulative. *Wetmore vs. Tracy*, (14 *Wend.* 250.)

The reference to § 45 is according to numbering in 1st Ed. R. S. The remedy does not appear to the compiler to be appropriate.

§ 126. [*Sec.* 105.] If the occupant to whom notice is given, shall (within five days) deny such encroachment, the commissioners, or some of them, shall apply to any justice of the peace of the county, for a precept directed to any constable of the town, to summon twelve freeholders thereof, to meet at a certain day and place to be specified in such precept, and not less than four days after the issuing thereof, to inquire into the premises. The constable to whom such precept shall be directed, shall give at least three days' notice to the commissioners of highways of the town, and to the occupant of the land, of the time and place at which such freeholders are to meet. [*Amended by* § 2 *of same chapter.*]

Proceed'gs if encroachment be denied.

13 John. 460.
3 Wend. 468.

The justice should not annex to the precept the list of jurors to be summoned. If he should do so, however, the objection should be made at the time, or it will be deemed to have been waived. *Mott vs. Commissioners of Rush*, (2 *Hill*, 473,) 11 *Pick. R.* 269.

One commissioner can enter the complaint. *Bronson vs Mann*, (13 *Johns. R.* 460.

§ 127. [*Sec.* 106.] On the day specified in the precept, the jury so summoned, shall be sworn by such justice, well and truly to inquire whether any such encroachment has been made, and by whom. Such witnesses as may be produced by either party shall also be sworn by such justice; and the jury shall hear the proofs and allegations which may be produced and submitted.

Ibid.

See Appendix, Nos. 62 *and* 63.

In *Pugsley vs Anderson*, 3 *Wend.* 470, the court decided under this section, that the justice, after issuing the precept for summoning the jury, had no duty to perform on the inquiry, except to swear the jury and the witnesses. He does not act judicially; he has no judgment to render, nor order of any kind to enter; if the costs are not paid in ten days, it is his duty to issue a warrant, not an execution, for its collection. He is, of course, to tax the costs, but he has not the custody of the certificate of the finding of the jury; and it seems to me he has no right to decide upon the qualification of the jurors.

Consequently a *certiorari* will not lie to a justice of the peace to bring up the proceedings had under this section.—*Ibid.*

If the jury should disagree, they may be discharged, and the justice can issue a new precept and another jury be summoned; so on until a jury do agree. *Fitch vs. Commissioners of Kirkland,* 22 *Wend.* 132. 2 *R. S.* 554, § 26, where the general rule is laid down.

Verdict how enforced.

§ 128. [*Sec.* 107.] If the jury find that any encroachment has been made, they shall make and subscribe a certificate in writing, stating the particulars of such encroachment, and by whom made; which shall be filed in the office of the town clerk. The occupant of the land, whether such encroachment shall have been made by him or by any former occupant, shall remove his fences within sixty days after the filing of such certificate, under the penalty provided in the one hundred and fourth section of this Title. He shall also pay the costs of such inquiry: and if the same shall not be paid within ten days, the justice shall issue a warrant for the collection thereof, in the manner provided in the forty-third section of this Title.

See Appendix, No. 64.

The certificate should be very particular; it is intended as a guide to the party who is convicted of the encroachment. In removing it, he is to conform to the certificate, which must, in the words of the statute, "state the particulars of the encroachment." *Fitch vs Commissioners of Kirkland,* (22 *Wend.* 132.) *Mott vs Commissioners of Rush,* (2 *Hill* 472.)

The breadth of the road originally intended, and the extent of the encroachment by the party upon that breadth, and the place or places where, ought to be specially stated, so that the party may be able to obey the order, and know when he had performed his duty. *Spier vs. Slade,* (9 *Johns. R.* 359.)

In an action brought to recover the penalty, this certificate of the finding of the jury is conclusive evidence of the fact of the existence of such encroachment. *Bronson vs. Mann*, (13 *Johns. R.* 460.) *Fleet vs. Young*, (7 *Wend.* 291.) *Pugsley vs. Anderson*, 3 *Wend.* 468.

The certificate partakes of a judicial character. *Cowen & Hill's Notes*, 1051.

§ 129. [*Sec.* 108.] If the jury find that no encroachment has been made, they shall so certify, and shall also ascertain and certify the damages which the then occupant shall have sustained by such proceeding; which, together with the cost thereof, shall be paid by the commissioners, and shall be a charge in their favor against the town by which they shall have been elected. Ibid.

See Appendix, No. 65.

This certificate is conclusive in favor of the person complained of, and may be used as a bar to subsequent proceedings for the same alleged encroachment. The party interested should see that the certificate is regularly made out, signed and filed, for his future protection.

§ 130. [*Sec.* 109.] No person shall be required to remove any fence under the preceding provisions of this Article, except between the first day of April and the first day of November in any year. When fences to be remov'd

§ 131. It shall be lawful for any person owning or occupying lands adjoining a highway or road, to construct a side-walk within such highway or road, and along the line of such land, and also to erect a railing thereto; and when a side-walk shall be so constructed, every person who shall ride or drive a horse or team upon it, shall forfeit the sum of one dollar, to the use of such owner or occupant, to be sued for in any court having cognizance thereof. [*Sec.* 1 *of chap.* 281 *of* 1836.] Penalty for injuring side walks.

Saving clause.

§ 132. This act shall not be so construed as to diminish, or in any wise interfere with the authority of the overseers of highways, or any other authority legally exercised over highways or roads. [*Sec.* 2 *of same chap.*]

Fallen trees to be removed.

§ 133. [*Sec.* 110.] If any tree shall fall or be fallen by any person from any enclosed land into any highway, any person may give notice to the occupant of the land from which such tree shall have fallen, to remove the same within two days. If such trees shall not be removed within that time, but shall continue in such highway, the occupant of the land shall forfeit the sum of fifty cents for every day thereafter until such tree shall be removed.

This notice may be verbal, and the penalty must be sued for by the commissioners of highways.

Penalty for falling trees.

§ 134. [*Sec.* 111.] In case any person shall cut down any tree on land not occupied by him, so that it shall fall into any highway, river or stream, unless by the order and consent of the occupant, the person so offending shall forfeit to such occupant the sum of one dollar for every tree so fallen, and the like sum for every day the same shall remain in such highway, river or stream.

For not removing from streams.

§ 135. [*Sec.* 112.] Whoever shall cut, or cause to be cut down, any tree, so that the same shall fall into any river or stream which now is or hereafter shall be declared a public highway, and shall not remove the same out of such river or stream within twenty-four hours thereafter, shall forfeit five dollars for every tree so cut down and left remaining.

The legislature have, from time to time, declared certain rivers and streams public highways, and reference must be had to the several acts. They would occupy too much space to enumerate them. All rivers where the tide ebbs and flows are a public highway, without an act of the legislature. 3 *Kent's Com.* 410, 6 *Cow.* 518, *Cowen's Treatise* 390.

§ 136. [*Sec.* 113.] No swinging or other gates shall be allowed on any public highway, laid out by virtue of this Title, or which has been heretofore laid out, other than such public highways as run through lands liable to be overflowed by the waters of the adjacent rivers or streams in such manner as to remove the fence thereon. Swinging gates.

§ 137. [*Sec.* 114.] Such gates shall be erected and kept in good repair by the overseers of highways of the town, at the proper costs and charges of the occupant of the land, for whose benefit the same shall be erected. How erected and preserved.

§ 138. [*Sec.* 115.] If more than one gate shall be erected, and the intermediate land between the gate, at the extremities of such lands, shall be in the occupation of more than one person benefitted by such gates, the whole charge of erecting and keeping the same in repair shall be borne by all the occupants benefitted thereby, in proportion to the extent of land each occupies adjoining the highway, between the gates at the extremities aforesaid. Expense.

§ 139. [*Sec.* 116.] The overseer of every road district in which such gates shall be, shall, on or before the first day of November in every year, make out and file with the town clerk, a statement of the charges incurred in the erection or repairing of such gates, with the name of the person bound to defray the same; which account shall be verified by the oath of such overseer. If more than one person is liable to defray Proceedings to collect.

such charges, the statement shall also contain an apportionment thereof between such persons, stating the amount to be paid by each.

See Appendix, No. 66.

Ibid.

§ 140. [*Sec.* 117.] The overseer shall, within ten days after filing the statement, demand of every person bound to pay such charges, or to contribute thereto, the sum due from him according to such statement; and if any person shall refuse or neglect to pay such moneys within six days after demand, it shall be the duty of the overseer to make complaint to a justice of the peace of the town, and the like proceedings shall be had for the recovery of such moneys, as in the recovery of fines for refusing or neglecting to work on the highways.

Gates to be closed, &c.

Penalty.

§ 141. [*Sec.* 118.] The commissioners of highways shall file an account of such gates in the town clerk's office; and if any person shall open any such gate, and shall not, immediately after having passed the same, close it, or shall wilfully or unnecessarily ride over any of the ground adjoining the road in which such gates shall be permitted, he shall forfeit to the party injured, treble damages.

Lands adjoining a public highway, remaining unenclosed, are considered as dedicated to the public use, and no action will lie by the owner against any person travelling over it. *Cleveland vs. Cleveland*, (12 *Wend.* 172.)

If passengers have been used, time out of mind, when the roads are bad, to go by outlets on the land adjoining to a highway, in an open field; and therefore, if they are sown, and the track is foundrious, passengers may go on the part sown. *Croke's Car.* 366, *Douglass*, 749.

A person traveling on a *public* highway, which has become foundrious and impassable, has a right to remove enough of the fences in the adjoining close, to enable him to pass around the obstruction, doing no unnecessary injury; but he becomes a trespasser if he tears away other fences, and tramples down the herbage in other parts of the close.

But the grantee of a *private* way, which has become foundrious and impassable, can not, without being a trespasser, go on the adjoining close, and thus pass around the obstruction. *Williams vs. Safford, 7 Bar. S. C.* 309.

ARTICLE SIXTH.

Of the erection, repairing and preservation of Bridges.

§ 142. [*Sec.* 119.] Whenever it shall appear to the board of supervisors of any county, that any one of the towns in such county would be unreasonably burthened by erecting or repairing any necessary bridge or bridges in such town, such board of supervisors shall cause such sum of money to be raised and levied upon the county, as will be sufficient to defray the expenses as they may deem proper; and such moneys, when collected, shall be paid to the commissioners of highways of the town in which the same are to be expended.

When bridges at expense of county.

§ 143. [*Sec.* 120.] No board of supervisors shall, under the last preceding section cause any sum exceeding one thousand dollars, to be levied and raised on any county in any one year.

Limit.

The act of April 18, 1838, page 314, authorizes the supervisors to raise "*such sum* of money as may be necessary to construct and repair bridges, and to prescribe the plan and in what manner the money shall be expended;" and to apportion this tax on the towns and wards.

§ 144. [*Sec.* 121.] In case the commissioners of highways of any town shall be dissatisfied with the determination of the board of supervisors of their county, touching an allowance for any such bridges, such determination shall, on the application of the commissioners, be reviewed by the court of sessions of the same county, whose order in the premises shall

Provision respecting allowance made by the supervisors.

be observed by every such board of supervisors. [*Amended by* § 15 *of chap.* 455 *of* 1847.]

This application may be made by the commissioners of any town in the county. The notice of the application should be served on the chairman of the board of supervisors; or, in case there is none, or he is absent, it may be served on the clerk of the board. 2 R S. p. 458 § 5. Although no time is prescribed, it is proper that the notice should be served ten days previous to the sitting of the next court.

Bridges how to be built and repaired between towns.

§ 145. Whenever any adjoining towns shall be liable to make and maintain any bridges over any stream dividing such towns, such bridges shall be built and repaired at the equal expense of said towns, without reference to the town lines. [*Sec.* 1 *of chap.* 225 *of* 1841.]

Joint contract to be entered into by commissioners.

§ 146. For the purpose of building and repairing such bridges, it shall be lawful for the commissioners of highways in said adjoining towns to enter into joint contracts, and such contracts may be enforced in law or equity against such commissioners jointly, the same as if entered into by individuals; and said commissioners may be proceeded against jointly for any neglect of duty in reference to such bridges. *Sec.* 2 *of same chap.*

Proceedings in case of neglect or refusal of either town.

§ 147. If the commissioners of highways of either of such towns, after reasonable notice in writing from the commissioners of highways of any other of such towns, shall neglect or refuse to build or repair any such bridge, it shall be lawful for the commissioners so giving such notice, to make or repair such bridges, and then to maintain a suit at law in their official capacity against said commissioners so neglecting or refusing to join in such making or repairing; and in such suit, the plaintiff shall be entitled to recover one-half of the expense of such building or repairing, with cost of suit and interest, without proving any contract. [*Sec.* 3 *of same chapter.*]

Judgment recovered to be a charge on town, how collected.

§ 148. Any judgment recovered against the commissioners of highways in their official capacity under the provisions of this act, shall be a charge on said town, and collected in the same manner as other town charges, except in cases where the court before which the judgment shall be recovered shall certify that the neglect or refusal of said commissioners was wilful and malicious, in which case said commissioners shall be personally liable for such judgment, and the same may be enforced against them in the same manner as against individuals. [*Sec.* 4 *of same chap.*]

Notice of fine, &c.

§ 149. [*Sec.* 122.] The commissioners of highways of each town may put up and maintain in conspicuous places, at each end of any bridge in such town, maintained at the public charge, and the length of whose chord is not less than twenty-five feet, a notice with the following words in large characters "One dollar fine for riding or driving on this bridge faster than a walk."

Penalty.

§ 150. [*Sec.* 123.] Whoever shall ride or drive faster than on a walk, over any bridge, upon which such notices shall have been placed, and shall then be, shall forfeit for every offence, the sum of one dollar.

This penalty is recoverable by the commissioners of highways of the town where the offence is committed. If the bridge is over a stream that divides two towns, and the offence is only on that part of the bridge in one town, that town can alone recover; if the offence is continued across the bridge, the commissioners of either town may prosecute, and a recovery by one town will be a bar to a suit commenced by the other.

Injuries to bridges.

§ 151. [*Sec.* 124.] Whoever shall injure any bridge maintained at the public charge, shall, for every offence, forfeit treble damages.

In addition to this penalty, it is enacted, 2 R. S. p. 695, § 30; "Every person who shall wilfully or maliciously destroy any public or "toll bridge, or turn-pike gate, shall, upon conviction, be adjudged guilty of a misdemeanor."

And, also, it is declared to be *arson* in the fourth deee to wilfully set on fire to or burn, in the day or night time, "any toll bridge, "or any other public bridge." 2 R. S. p. 667, § 7.

The following act was passed May 25, 1841, on the subject of bridges:

AN ACT *relating to the joint liability of Commissioners of Highways.* Chap. 225, p. 207.

§ 1. "Whenever any adjoining towns shall be liable to make "and maintain any bridges over any stream dividing such towns, "such bridges shall be built and repaired at the equal expense of "said towns, without reference to the town lines.

§ 2. "For the purpose of building and repairing such bridges, it "shall be lawful for the commissioners of highways in said adjoining "towns to enter into joint contracts; and such contracts may be en- "forced in law or equity against such commissioners jointly, the "same as if entered into by individuals; and said commissioners "may be proceeded against jointly for any neglect of duty in refer- "ence to such bridges.

§ 3. "If the commissioners of highways of either of such towns "after reasonable notice in writing from the commissioners of high- "ways of any other such towns, shall neglect or refuse to build or "repair any such bridge, it shall be lawful for the commissioners so "giving such notice, to make or repair such bridges, and then to "maintain a suit at law in their official capacity against said commis- "sioners so neglecting or refusing to join in such making or repair- "ing; and, in such suit, the plaintiff shall be entitled to recover one- "half of the expense of such building or repairing, with costs of suit "and interest, without proving any contract.

§ 4. "Any judgment recovered against the commissioners of high- "ways in their official capacity under the provisions of this act, shall "be a charge on said town, and collected in the same manner as other "town charges, except in cases where the court before which the "judgment shall be recovered shall certify that the neglect or refu- "sal of said commissioners was wilful and malicious, in which case "said commissioners shall be personally liable for such judgment, "and the same may be enforced against them in the same manner as "against individuals."

Canal Bridges.—The provisions in relation to canal bridges will be found convenient for those officers who have the care of roads and bridges in those parts of the State intersected by canals. The following provisions will be found in 1 R. S. page 247, § 174, 175, 176.

§ 174. "In all cases where a new road or public highway shall be laid out by legal authority, in such direction as to cross the line of any canal, and in such manner as to require the erection of a new bridge over the canal, for the accommodation of the road, such bridge shall be so constructed and forever maintained, at the expense of the town in which it shall be situate.

§ 175. "No bridge shall be constructed across any canal, without first obtaining for the model and location thereof, the consent in writing of one of the canal commissioners, or of a superintendent of repairs, on that line of the canal which is intersected by the road.

§ 176. 'Every person who shall undertake to construct or to locate such bridge without such consent, and shall proceed therein so far as to place any materials for that purpose on either bank of the canal, or on the bottom thereof, shall forfeit the sum of fifty dollars; and each of the commissioners, superintendents, or engineers, shall be authorized to remove all such materials, as soon as they are discovered, wholly without the banks of the canal."

Another provision was also made in relation to bridges, by an act passed April 18, 1838.

AN ACT *in relation to Turnpike Roads and Toll-Bridges.*

Passed April 18, 1838.

The People of the State of New York, represented in Senate and Assembly, do enact as follows:

§ 1. "Whenever any turnpike corporation shall become dissolved or the road discontinued, its road shall become a public highway, and be subject to all the legal provisions regulating highways.

§ "It shall be lawful for any corporation or individual owning a toll-bridge, to put up at each end thereof, in a conspicuous place, a notice in the following words in large characters: 'One dollar fine for riding or driving faster than a walk on this bridge;' and, during the continuance of such notice, any person who shall ride or drive faster than a walk on such bridge, shall forfeit the sum of one dollar, to be sued for in the name of the corporation or person or persons owning such bridge, and to be recovered with costs of suit.

§ 3 "Whenever a corporation owning a toll-bridge shall become dissolved, such bridge shall be left without waste or damage, and shall be a public highway."

It seems that where a bridge is built by an individual over a natural stream, for his own benefit, if the bridge be of *public utility*, and is used by the public, the public are bound to keep it in repair; but not so when the necessity for the bridge is created by the individual. *Dygert vs. Schenck,* 23 *Wend.* 446, *and cases there cited.*

§ 152. It shall be lawful for any corporation or individual owning a toll-bridge, to put up at each end thereof, in a conspicuous place, a notice in the following words in large characters: "One dollar fine for riding or driving faster than a walk on this

Notice at toll bridges

bridge;" and during the continuance of such notice, any person who shall ride or drive faster than a walk on such bridge, shall forfeit the sum of one dollar, to be sued for in the name of the corporation or person or persons owning such bridge, and to be recovered with costs of suit. [*Sec.* 2 *of chap.* 262 *of* 1838.]

When to become part of highway.

§ 153. Whenever a corporation owning a toll-bridge shall become dissolved, such bridge shall be left without waste or damage, and shall be a public highway. [*Sec.* 3 *of same chapter.*]

AN ACT *to provide for the iucorporation of Bridge Companies.*

Passed April 11, 1848.

The People of the State of New York, represented in Senate and Assembly, do enact as follows:

§ 1. Any number of persons not less than five, may be formed into a corporation, for the purpose of constructing and owning a bridge across any stream of water, as hereinafter provided, upon complying with the following requisitions:

1. They shall severally subscribe articles of association, in which shall be set forth the name of the corporation, the number of years the same is to continue, which shall not exceed fifty years; the amount of the capital stock of the corporation, which shall be divided into shares of twenty-five dollars each, the number of directors and their names, who shall manage the concerns of the corporation for the first year, and until others are elected; the location of such bridge, and the plan thereof:

2. Each subscriber to such articles of association shall subscribe thereto his name and place of residence, and the number of shares of stock taken by him in such corporation.

3. Whenever one-fourth part of the amount of the capital stock, specified in the articles of association, shall have been subscribed, and on complying with the provisions of the next section, such articles may be filed in the office of the state engineer and surveyor, and clerk of the county or counties in which the bridge is built; and thereupon the persons who have subscribed the articles of association as aforesaid, and such other persons as shall become stockholders in such company, and their successors, shall be a body corporate, by the name specified in such articles of association, and shall possess the powers and privileges, and be subject to the provisions of titles three and four of chapter eighteen of the Revised Statutes, so far as those provisions are consistent with the provisions of this act.

§ 2. All the stockholders of every company incorporated under this act, shall be severally and individually liable, to an amount equal to the amount of the capital stock held by them respectively, to the creditors of such company, for all the debts contracted by the directors or agents of such company for its use, until the whole amount of the capital stock fixed and limited by such company is paid in, and a certificate thereof filed in the offices aforesaid, and the whole capital stock paid in, shall be one-half thereof within one year, and the other half thereof within two years from the time of the incorporation of such company, and if not so paid in, such corporation shall be dissolved. If the directors of any corporation formed under this act shall contract debts for the company, exceeding in the aggregate the amount of the capital stock, they shall be personally liable for all the debts of the corporation.

§ 3. Such articles of association shall not be filed as aforesaid, until five per cent on one-fourth of the amount of the stock of such company fixed as aforesaid shall have been actually paid in, in good faith, to the directors named in such articles of association, in cash, nor until there shall be endorsed thereon, or annexed thereto, an affidavit made by at least three of the directors named in such articles of association, that the amount of stock required by the first section of this act to be subscribed, has been subscribed, and that five per cent on the amount has been actually paid in as aforesaid.

§ 4. A copy of such articles of association filed in pursuance of this act, with a copy of such affidavit endorsed thereon or annexed thereto, and certified to be a copy by the proper officer, shall, in all courts and places be presumptive evidence of the facts therein contained.

§ 5. The business and property of every such corporation shall be managed and conducted by a board of directors, consisting of not less than five nor more than nine, who shall be chosen, except those for the first year, at such place within a county in which the bridge of such corporation or some part thereof shall be located, as shall be prescribed by the by-laws thereof. The directors shall give notice of every such election, previous to the holding thereof, by publishing the same once in each week for four successive weeks, in a public newspaper, published in each county in which such bridge or any part thereof, shall be located, and if in any such county no such paper shall be published, such notice shall be published in some county adjoining such last mentioned county. All elections of directors shall be by ballot and by a majority of all votes given thereat; and every stockholder being a citizen of the United States and attending in person or by proxy, shall be entitled to one vote for each share of stock which he shall have owned absolutely, or as executor, administrator or guardian, for thirty days previous to such election. No person shall be a director unless he shall be a stockholder, owning at least four shares of stock, absolutely in his own right or as executor, administrator or guardian, and entitled to vote at the election at which he shall be chosen, nor unless he shall be a citizen of this state ; and a majority of the directors shall, at the time of their election be residents of the county or counties in which such bridge shall be located. Whenever any vacancy shall happen in the board of directors, it shall be supplied until the next election by the remaining directors. The directors of every such company shall be elected in the same month in each and every year, and such election after the first, shall be held on the first Tuesday of such month, and the directors chosen at any election shall hold their offices, to and including the Tuesday next after that appointed by law for holding the election next succeeding that at which they were chosen. If an election of directors shall not be held on the day prescribed by this act for holding the same, the directors in office on that day shall hold their offices until their successors shall be elected, but after the expiration of their regular term of office as prescribed by this section, they shall be incapable of doing any act, as such directors, except such as may be necessary to give effect to an election of directors. The provisions of the second article of the second title of the eighteenth chapter of the first part of the Revised Statutes, shall apply to every corporation formed under this act, so far as such provisions shall be consistent with the provisions of this act.

§ 6. When any bridge corporation shall be desirous of constructing a bridge or any part thereof, in any county, it shall apply to the board of supervisors of such county at the annual or any special meeting thereof, for authority to construct such bridge; of which application, such corporation shall give notice, by publishing the same in at least one public newspaper in such county, or if no newspaper is published therein, then in an adjoining county, once in each week for six weeks successively, previous to the time of presenting such application to such board, specifying such time and the location of such proposed bridge. If the place of the location of such bridge shall be situated in more than one county. such application shall be made to the board of supervisors of every such county. Such application shall also specify the length and breadth of such bridge ; and the notice of such application shall set forth all the particulars required to be specified in such application. Upon the hearing of the said application, all persons residing in such county or interested in such application, may appear and be heard in respect thereto. Such board may take testimony in respect to such application, or may suthorize it to be taken by a county judge or justice of the peace of such county; and it may adjourn the hearing from time to time. A copy of the articles of association of such corporation certified by the state engineer and surveyor, or by the clerk where such articles are filed, shall be attached to and filed with such application. No such corporation shall be authorized to bridge any stream, in any manner that will prevent or endanger the passage of any raft of forty-five feet in width, or any ark where the same is navigated by rafts or arks.

§ 7. If after hearing such application such board shall be of opinion that the public interests will be promoted by the construction of such bridge on the proposed site, it may, if a majority of all the members elected to such board, shall assent thereto, by an order to be entered in its minutes, authorize such company to construct such bridge, as shall have been specified in the application which shall be particularly described in such order. Such corporation shall cause a copy of such order certified by the clerk of such board, with such application, to be recorded in the clerk's office of such county, before it shall proceed to do any act by virtue thereof; and such board shall cause such application when it shall have finally acted on the same, to be filed at the expense of the corporation, with all the other papers relating thereto, or to the proceedings of said board thereon, in the office of the clerk of the county in which it shall have been made. Any corporation formed under this act, may use in such manner as such board shall prescribe, so much of any public highway on either side of any stream, as may be necessary for the construction and maintenance of such bridge and toll houses.

§ 8. In case any bridge shall be constructed under the provisions of this act, over any stream navigable by rafts, it shall be the duty of the corporations constructing such bridge, at all times to keep the channel of said stream both above and below said bridge, free and clear from all deposits in any wise prejudicial to the navigation thereof, which may be formed or occasioned by the erection of such bridge,

§ 9. Any corporation organized under the provisions of this act, which shall construct any bridge over any stream, navigable by rafts as hereinbefore provided, shall be liable to pay all persons who may be unnecessarily or unreasonably hindered or delayed in passing such bridge, all damages which they shall sustain thereby, to be recovered with costs of suit.

§ 10. Every bridge constructed by virtue of this act, shall be built with a good and substantial railing or siding, at least four and a half feet high. Whenever such bridge shall be completed and a certficate signed by the county judge of the county in which such bridge is situated, or if such bridge shall be located in more than one county, by the county judge of each of such counties, and such certificate filed in the office of the clerk of such county or of each of said counties, if such bridge shall be located in more than one county, that such bridge is constructed and completed in a manner safe and convenient for the public use, the directors may erect a toll gate at such bridge, and demand and receive such sum as shall be from time to time prescribed by the supervisors of the county or counties where the bridge is located.

§ 11. No tolls shall be collected for crossing any bridge constructed by any corporation formed under this act, from any person going to or from public worship, or to or from a funeral, or to or from shool, or to or from a town meeting or election, at which he is entitled to vote, for the purpose of giving such vote, and returning therefrom; or to or from a military parade which he is, by law required to attend, or to or from any court which he shall be required to attend, as a juror or witness, or to or from his legally required work upon any public highway.

§ 12 .The directors of any incorporation formed under this act,may require payment from the stockholders of the sums subscribed to the capital stock, at such times and in such proportions and on such conditions as they shall see fit, under the penalty of the forfeiture of their stock, and all previous payments thereon; and they shall give notice of the payments thus required, and of the place and time when and where the same are to be made, at least thirty days previous to the time fixed for the payment of the same, for the time and in the manner herein prescribed for giving notice of the election of directors, and by sending such notice to such stockholder, by mail, directed to him at his usual place of residence.

§ 13. The shares of any corporation formed under this act, shall be deemed personal property, and may be transferred in such manner as shall be prescribed by the by-laws of such corporation; and the directors of every such corporation may, at any time, with the consent of a majority in amount of the stockholders in snch corporation provide for such increase of the capital stock thereof, as may be necessary for the completion or reconstruction of such bridge, and the certificate of the amount of any such increase, within thirty days thereafter, shall be filed in the offices of the state engineer and surveyor, and the clerk or clerks of the county or counties in which such bridge is located, which certificate shall be autheuticated by the signatures and oaths of a majority of said directors.

§ 14. So much of any such bridge or toll houses constructed by virtue of this act, as shall be within any town, city or village, shall be liable to taxation in such town, city or village, as real estate.

§ 15. Every company incorporated under this act, shall cease to be a body corporate:

1. If within two years from the filing of their articles of association, they shall not have commenced the construction of theirbridge and actually expended thereon at least ten per cent. of the capital stock of such company; or,

2. If within five years from the filing of such articles of association such bridge shall not be completed according to the provisions of this act; or,

3. If in case the bridge of such company shall be destroyed, it shall not be reconstructed within three years theaeafter.

§ 16. It shall be the duty of the president and secretary of every corporation formed under this act, to report annually to the state engineer and surveyor, and the county clerk where the papers are filed, under oath, the costs of their bridge, the amount of all money expended, the amount of their capital stock, and how much paid in, and how much actually expended, the amount received during the year for tolls, and from all other sources, stating each separately, the amount of dividends made, and the amount of indebtedness of such company, specifying the object for which the indebtedness accrued; and such other particulars in respect to the business affairs of such corporation, as the said state engineer and surveyor, or the legislature, or either branch thereof require to be so reported.

§ 17. When any bridge may be in process of construction by private subscriptions at the time of the passage of this act, the subscribers may organize into a corporation pursuant to the provisions of this act, with the same power and privileges as if such bridge had not been so commenced.

§ 18. All companies formed under this act shall at all times be subject to visitation and examination by an officer or agent, in pursuance of law, or by the legislature, or by a committee, appointed by either house thereof; and the courts of this state shall have the same jurisdiction over such corporations and their officers as over those created by special acts.

§ 19. Every report required to be made by the 16th section of this act, shall be made in the month of January in each year, and shall show in respect to the particulars required therein to be set forth, the affairs and business of the corporation, making the same at the close of the year, ending on the thirty first day of December, next preceding the time of making the same, and shall be published in the nearest newspaper four weeks, and every corporation formed under this act, which shall neglect to make such report as thereby required, shall forfeit to the people of this state for every such neglect, the sum of two hundred dollars, and for every week such corporation shall neglect to make such report after the expiration of the time, within which it is required as aforesaid to make the same, it shall forfeit as aforesaid the further sum of fifty dollars. The state engineer and surveyor shall report to the attorney-general every such forfeiture, by whom the same shall be sued for and recovered with the costs in the name of the people; and the certificate of the said state engineer and surveyor of any such neglect shall be presumptive evidence thereof, and if any such river, water course or lake, now so navigable, shall hereafter be rendered navigable up stream by vessels or steamboats, power to require such bridge to be altered or removed is reserved to the legislature.

§ 20. Nothing in this act shall be construed so as to authorize the bridging of any river or water course where the tide ebbs and flows, or any water used for a harbor, any lake, river or water, which is navigable by sail vessels or steamboats, nor the construction of any bridge within the limits prescribed by any existing law for the erection or maintenance of any other bridge.

§ 21. Any existing corporation having for its object the construction and maintenance of any bridge whose charter shall expire, may be continued as such corporation by complying with the provisions of this act, so far as the same are applicable to them, with the consent of the supervisors of the county or counties in which their bridge is located, to be obtained on application to them as herein before provided.

§ 22. This act shall take effect immediately.

ARTICLE SEVENTH.

Miscellaneous Provisions of a General Nature.

Two commissioners may act.

§ 154. [*Sec.* 125.] Any two commissioners of highways, of any town, may make any order, in execution of the powers conferred in this Title; provided it shall appear in the order filed by them, that all the commissioners of highways of the town met and deliberated on the subject embraced in such order, or were duly notified to attend a meeting of the commissioners, for the purpose of deliberating thereon.

This section is in accordance with the general provision of the statute, 2 *R. S., page* 555, § 27.

The order made by the commissioners should recite particularly that they all met or were duly notified, or it will be held to be void.

In an order made under § 103 to remove encroachments, it recited "at which meeting *F. B., two of the* commissioners " of *said town were present, and G. G. having been duly notified* did not attend, (BUT WAS) deemed and adjudged by the two commissioners (TO BE) present." This differs from words of the statute, which gives the form. The recital stops with saying *he was duly notified;* but does not add for what purpose. The statute requires that the notice should be given, for *the purpose of deliberating* on the subject of *the specific encroachment,* or other subject, whatever it may be; and that the order should expressly state the purpose of the notice. Where the statute prescribes the form, the very words of the order or other summary proceedings, those words must be used, at least as far as they can be applied to the nature and exigency of the particular proceeding. *Fitch vs. coms. of Kirkland.* (22 *Wend.* 135,) *and cases there cited.*

Consent to railroads crossing highways.

§ 155. Whenever any association or individual shall construct a rail-road upon land purchased for that purpose, on a route which shall cross any road or other public highway, it shall be lawful for the commissioners of highways having the supervision therof, to give a written consent that such rail-road may be constructed across or on such road or other public highway; and thereafter such association or individual shall be authorized to construct and use such rail-road across or on such roads or other highways, as the commissioners aforesaid shall have permitted;

but any public highway thus intersected or crossed by a rail-road, shall be so restored to its former state as not to have impaired its usefulness.

If the Rail-Road Company purchase the land for their track on each side of the highway, of course they become the owners of the fee in the highway; and by getting this consent from the commissioners, they have full right to lay the track. They must acquire the fee in the road before they can lay their track, either by gift, or by making compensation to the owners. As the public only have an easement or right of passage in the highway, subject to this easement, the rights and interests of the owner of the fee remain unimpaired. *Trustees of the Presbyterian Church in Waterloo, vs. Auburn and Rochester Rail-Road Company,* (3 *Hill*, 567.) 2 *Strange* 1004. 1 *Wils.* 110, 111. 6 *Mass.* 454. 6 *East.* 154. 15 *Johns.* 447. 12 *Wend.* 98.

The legislature cannot authorize the construction of a rail-road across a highway without making a provision for the compensation of the owner of the fee.

Trees, to whom they belong.

§ 156. [*Sec.* 126.] All trees standing or lying on any land over which any highway shall be laid out, shall be for the proper use of the owner or occupant of such land, except such of them as may be requisite to make or repair the highways or bridges on the same land.

Trees may be planted.

§ 157. [*Sec.* 127.] Any person owning land adjoining any highway not less than three rods wide, may plant or set out trees on the side of such highway contiguous to his land; which trees shall be set in regular rows, at a distance of at least six feet from each other. Whoever shall cut down, destroy or injure any tree that has been or shall be so planted or set out, shall be liable in damages to the owner of such adjoining land.

Penalty for injuring.

These sections recognize the fundamental principle of the common law in relation to roads, that the *grant* or *laying out* of a highway gives only a right of way to the public, and that the fee or right of soil remains in the original owner. 2 *Johns.* 357. 1 *Wend.* 262. 20. *Johns.* 743. 15 *Johns.* 447, 452.

The fee of the highway belongs to the owner of the adjoining ground, and the public have only a right of passage. It is but a *servitude* or easement; and trespass will lie for any exclusive appropriation of the soil. 1 *Burr.* 143. 2 *Strange*, 1004. 1 *Wills.* 107. 6 *East.* 154.

He retains the exclusive right of all mines, quarries, springs of water, timber and earth, for every purpose not incompatible with the public right of way. *See Cowen's Treatise* 371, 818. 6 *Mass. R.* 454. 2 *Johns.* 363. *Woolrych on Ways*, 5.

In the case of *Fish vs. Mayor of Rochester*, (6 *Paige*, 272,) the Chancellor says: "I believe it is the common practice of public officers having the care of public roads, to take the materials which are removed from one part of the highway under their direction, in improving the road at that point, and depositing them wherever they may be wanted for the repair or improvement of the highway in other places, even beyond the boundaries of the lands opposite to where the materials are taken. The only restriction that I am aware of, is that contained in § 126, which gives to the owner or occupant of land, over which a highway is laid out. the use of the trees thereon, except such as are requisite to make or repair highways or bridges on the same land. This legislative restriction of the right to use trees, or limiting it to repairs of the road within the same lands, appears to be founded upon the supposition that without such restriction they might be used to repair or improve other parts of the highway; and as the restriction is confined to trees, other materials may be used beyond the bounds of the land from which they are taken. Trees, when cut down, may be left without much inconvenience, by the side of the highway, until the owner of the land through which the road runs, has a reasonable time to remove them. But when it is necessary to excavate earth or ground, it must be removed entirely or the work cannot be completed, and must be done anew. And it cannot be deposited upon lands adjacent to the highway, without the consent of the owners of such land. When in a state of removal, therefore, it does not appear to be susceptible of any distinct ownership except in those under whose authority or by whom such removal is made."

Penalty for injur- mile-board

§ 158. [*Sec.* 128.] Whoever shall destroy, remove, injure or deface any mile-board or mile-stone, erected on any highway, shall forfeit for every offence, the sum of ten dollars; he shall also be deemed guilty of a misdemeanor, and, on conviction, shall be fined not exceeding fifty dollars, or imprisoned not exceeding three months, at the discretion of the court.

Do. as to guide-posts.

§ 159. [*Sec.* 129.] Whoever shall injure or deface any description affixed to a guide-post, erected on any highway, or destroy or injure any such guide-post, shall be liable to all the penalties provided in the last preceding section.

§ 160. [*Sec.* 130.] Whoever shall injure any highway, by obstructing or diverting any creek, water-course or sluice, or by drawing logs or timber on the surface of any road or bridge, or by any other act, shall, for every such offence, forfeit treble damages. Injuries to road.

Commissioners cannot by virtue of their offices bring suits to recover damages against individuals or corporations for illegally entering upon and taking possession of the highways or bridges of their town. (*Cornell & Clark vs. Town of Guilford,*) 1 *Denio*, 510. In that case a turnpike company entered upon and took possession of a highway and bridge in Guilford, and the commissioners, in pursuance of a vote of the town, at a town meeting, prosecuted the company, and were unsuccessful, and were compelled to pay a large amount of costs and expenses of the litigation. The commissioners sued the town, and it was decided that the town had no power, by resolution or otherwise, to authorize the commissioners to bring an action for such injuries. Such a resolution would not bind the town, and, consequently, that they could not recover of the town.

This is a penalty recoverable by the commissioners. They cannot maintain an action in the case for damages done to the road; the remedy is by indictment. summary abatement, or action for the penalty. *Cornell & Clarke vs. Butternuts Turnpike Co.* (25 *Wend.* 365.)

§ 161. [*Sec.* 131.] All penalties or forfeitures given in this Title, and not otherwise specially provided for, shall be recovered by the commissioners of highways of the town in which the offence shall be committed; and, when recovered, shall be applied by them in improving the roads and bridges in such town. Penalties how re-

They cannot maintain an action in their official name or title, but must use their individual name, annexing their official title. Coms. of H. of Cortlandville vs. Peck, (5 *Hill*, 215.) Supervisor of Galway vs. Stimson, (4 *Hill*, 136.) A suit should be commenced as follows: "A. B. and C., Commissioners of highways of the town of Galway, in the county of Saratoga, plaintiffs, &c."

In general, all public officers, though not expressly authorized by

Statute, have a capacity to sue commensurate with their public trusts and duties. *Overseers of Pittstown vs. Overseers of Plattsburg*, (18 *Johns.* 407.) *Todd vs. Birdsall*, 1 *Cow.* 260, *and notes pp.* 261—4. 3 *Wend.* 193. *Silver vs. Cumings*, 7 *Wend.* 181. *Avery vs. Slack*, 19 *Wend.* 50.

Commissioners of highways cannot by virtue of their office bring suits to recover damages against individuals or corporations for illegally entering upon and taking possession of the public highways or bridges of their town. *Cornell vs. the town of Guilford*, (1 *Denio* 510.)

The electors of a town in town meeting have no power by resolution or otherwise, to authorize commissioners of highways to bring an action in their own names, or in their name of office, for such injuries. Such a resolution, if passed at town meeting, would not bind the town. Where a cause of action exists in behalf of a town, and no officer is by statute authorized to prosecute for such cause of action, the town meeting may direct such action to be brought, and may appoint an agent to prosecute it, but such suit must be brought in the name of the town. Per Jewett, J.

Where the electors of a town, in their town meeting, directed the commissioners of highways to prosecute a turnpike company, for entering upon and taking possession of a public highway and bridge in that town, and the commissioners accordingly brought a suit for that cause of action, and had judgment against them; *held*, that they could not sustain an action against the town to be reimbursed their costs and expenses, or the costs recovered against them in that suit. Ibid.

Extent of this Title.

§ 162. [*Sec.* 132.] The provisions of this Title shall be construed to extend to all parts of the state, except where special provisions, inconsistent therewith, have been or shall be made by law, in relation to particular counties, cities, villages or towns.

§ 163. The first Title of the sixteenth Chapter of Part First of the Revised Statutes, shall, immediately on the passing of this act, be in force in the county of Richmond.

AN ACT *to provide for the incorporation and regulation of Telegraph Companies.*

Passed April 12, 1848.

§ 3. "Such association (Telegraph) is authorized to construct lines of telegraph along and upon any of the public roads and highways, or across any of the waters within the limits of this state, by the erection of the necessary fixtures, including posts, piers or abutments, for sustaining the cords or wires of such lines; provided the same shall not be so constructed as to incommode the public use of said roads or highways, or injuriously interrupt the navigation of said waters; nor shall this act be so construed as to authorize the construction of any bridge across any of the waters of this state."

It is very questionable whether the legislature could pass a law allowing a company to place erections in highways, to the injury of the owner of the soil, without *his consent*, or without making provisions for his compensation.

AN ACT *to authorize the formation of Railroad Corporations, and to regulate the same.*

Passed April 2, 1851.

§ 24. Whenever the track of a railroad, constructed by a company formed under this act, shall cross a railroad, a highway, turnpike or plank road, such highway, turnpike or plank road, may be carried under or over the track, as may be found most expedient; and in cases where an embankment or cutting shall make a change in the line of such highway, turnpike or plank road, desirable, with a view to more easy ascent or descent, the said company may take such additional lands for the construction of such road, highway, turnpike, or plank road, on such new line, as may be deemed requisite by the directors. Unless the lands so taken shall be purchased for the purposes aforesaid, compensation therefor shall be ascertained in the manner prescribed in this act for acquiring title to real estate, and duly made by said corporations to the owners and persons interested in such lands. The same, when so taken, shall become part of such intersecting highway, turnpike, or plank road, in such manner and by such tenure, as the adjacent parts of the same highway, turnpike or plank road, may be held for highway purposes.

§ 28. Sub. 5. To construct their road across, along or upon any stream of water, water course, street, highway, plank road, turnpike or canal, which the route of its road shall intersect or touch; but the company shall restore the stream or watercourse, street, highway, plank road, and turnpike they intersected or touched, to its former state, or to such state as not unnecessarily to have impaired its usefulness. Every company formed under this act, shall be subject to the power vested in the canal commissioners, by the seventeenth section of chapter two hundred and seventy-six of the session laws of eighteen hundred and thirty-four. Nothing in this act contained shall be construed to authorize the erection of any bridge, or any other obstructions across, in, or over any stream or lake navigated by steam or sail boats, at the place where any bridge or other obstructions may be proposed to be placed; nor to authorize the construction of any railroad not already located in, upon or across any streets in any city, without the assent of the corporation of such city.

TITLE II.

OF THE REGULATION OF FERRIES.

Liceses by whom granted.

§ 1. The court of common pleas in each of the counties of this state, shall grant licenses for keeping ferries in their respective counties, to as many suitable persons as they may think proper; which licenses shall continue in force for a term to be fixed by the court, not exceeding three years.

A ferry is a liberty, by license given, to have a boat for passage upon a river for carriage of horses and men, and goods, for a reasonable toll. *Cro El.* 591. *Woolrych on Ways*, 217.

The owner of a ferry cannot use the land on the other side of the stream, (unless he is himself the owner,) for the purpose of embarking and disembarking passengers. 3 *Kent's Com.* 421 ; *Peter vs. Lendall*, 6 *Barn. and Cress.*, 703. *See also Chamhers vs. Furry*, 1 *Yeates*, 167; *Cooper vs. Smith*, 9 *Serg. and Rawl.*, 26; *Saville*, 11 *pl.* 29.

If the land on the opposite side is a public highway, the owner of the ferry may use it.

The privilege of establishing a ferry and taking toll for the use of the same as a *franchise;* and the public have an interest in the same; and the owners of the franchise are liable to answer in damages, if they should refuse to transport an individual without any reasonable excuse, upon being paid or tendered the usual rate of fare. 3 *Kent's Com.* 458; *Beekman vs. Saratogo and Schenectady R. R. Co.*, 3 *Paige*, 45; *Paine vs. Patrick*, 3 *Mod. Rep.* 287–294; *Story J. in Charles River Bridge vs. Warren Bride*, 11 *Peter's Rep.* 639.

As there is now no court of common pleas, no such license can be obtained short of an application to the legislature. By the code, § 30, sub. 10, the power was conferred on the county court, until Januarry 1, 1850—that has now expired, and no court can now grant such license.

Notice of an application to the court for a license to establish a ferry, need not be given to all who claim a right to the ferry; nor even to all those who have obtained a license from another court for a ferry at the same place. All that is required, where the applicant is not the owner of the land through which the highway adjoining to the ferry runs, is that the person applying for a license shall give notice of the application, to the owners of such land. When a bill was filed, by persons claiming the exclusive right to a ferry, to obtain a decree restraining the defendant from keeping a ferry at the same place, and such bill alleged that the defendant had established a ferry there, in violation of the rights of the complainants; and that he was using the same in pursuance of a pretended license, from some court or person, but that if any license had been granted to the defendant, the same was granted in fraud of the complainant's rights, *and without any legal notice to them;* but such bill contained no allegation that the complainants were the owners of the land through which the road adjoining the ferry, ran. Held that the avement, as to the want of notice to the complainants, was not material for any of the purposes of the suit; and that the defendant was not bound to answer it. *Wiswell* vs. *Wundell*, 3 *Bar. Chan. Rep.* 312.

§ 2. No such license shall be granted to any person other than the owner of the land through which the highway adjoining to the ferry shall run, unless such owner shall neglect to apply for such license, after notice as hereinafter provided. To whom.

§ 3. Whenever application for a ferry shall be made by any person other than such owner, the court shall not grant a license to such applicant, unless proof shall be made, that the applicant caused notice in writing to be given to such owner, at least eight days before the sitting of the court, of his intention to make such application.

§ 4. Every person applying for such license shall, before the same be granted, enter into a recognizance to the people of this state, in open court, in the sum of one hundred dollars, faithfully to keep Recognizance.

and attend such ferry, with such and so many sufficient and safe boats, and so many men to work the same, as shall be deemed necessary, together with sufficient implements for said ferry, during the several hours in each day, and at such several rates as the court granting the license shall, from time to time, order and direct; which recognizance shall be forthwith filed with clerk of the county.

To be entered.

§ 5. Every license so granted shall be entered in in the book of minutes of the court by the clerk; and a copy thereof, attested by him, shall be delivered to the person licensed.

The clerk will enter the recognizance in the minutes of the common pleas, in the same manner as in criminal cases in the sessions, except the condition of the recognizance, which should be in the words of section 4.

Effect of certain licenses.

§ 6. Whenever the waters over which any ferry may be used shall divide two counties, a license obtained in either of the said counties shall be sufficient to authorize the person obtaining the same, to transport goods, wares, and merchandise, to and from either side of said waters.

In 11 *Wend.* 590, *The People vs. Babcock*, the question was raised, whether the common pleas of Niagara had power to grant licenses to keep *ferries* on the Niagara river, although the jurisdiction of the state extends only to the centre of the river; and it was decided that they had such power; and that persons keeping a ferry on that river without license will be subject to punishment as for a misdemeanor. *See Gibbons vs. Ogden*, 9 *Wheaton*, 1.

Penalty for misconduct.

§ 7. Every person who shall violate the condition of such recognizance shall be considered guilty of a misdemeanor; and, on conviction, shall be subject to such fine as the court may adjudge, not exceeding twenty-five dollars for each offence; and, on the proof of such conviction, the court of common pleas shall direct the recognizance entered into by such person to be estreated for the use of the people of this state.

Penalty for ferrying without license.

§ 8. If any person (except within the counties of Essex and Clinton, the counties of Orange, Rockland and Westchester, and the counties in the first Senate district,) shall use any ferry for transporting across any river, stream or lake, any person, or any goods, chattels, or effects, for profit or hire, unless authorized in the manner directed in this Title, such person shall be considered guilty of a misdemeanor; and, on conviction, shall be subject to such fine, for the use of the county, as the court may adjudge, not exceeding twenty-five dollars for each offence.

A party having a right of ferry granted cannot maintain an action for disturbance of his right. His only remedy is for the penalty under the statute. *Almy vs. Harris*, 5 *Johns. R.* 175.

Proceedings.

§ 9. Where any such offence shall be committed on waters dividing two counties, the person so offending may be proceeded against in each of said counties; but the fine to be imposed shall not exceed twelve dollars and fifty cents in each case.

Limitation of this Title.

§ 10. Nothing in this Title contained shall affect or alter the ferries granted by charter to the corporation of Albany and Hudson, or alter or impair any grants made by this state, or any legal right or privilege whatever, belonging to any individual or corporation, by virtue of any law of this state, or otherwise.

On the subject of ferries, see 2 *Hilliard on Real Prop.* 45. 11 *Peters*, 539. 3 *Kent*, 458.

Every commissioner of highways hereafter to be elected or appointed, shall, before entering upon his duties, and within ten days after notice of his election or appointment, execute to the supervisor of his town, a bond with two sureties, to be approved by the supervisor by an endorsement thereon, and filed with him, in the penal sum of one thousand dollars, conditioned that he will faithfully discharge his duties as such commissioner, and within ten days after the expiration of his term of office, pay over to his successor what money may be remaining in his hands as such commissioner, and render to such successor a true account of all moneys received and paid out by him as such commissioner. [*Sec.* 3 *of chap.* 180 *of* 1845.]

A bond taken in the names of the commissioners of highways of a town *virtute officii*, for the benefit of the town in its corporate capacity, and intended to relieve the taxable inhabitants of the town from the payment of a tax for a public improvement, *viz:* the extension and opening of a public highway—cannot be enforced against the obligors; the commissioners having no authority to take a bond of that nature, and the general policy of the law forbidding such a transaction. *Webb. vs. Albertson,* 4 *Bar. S. C. Rep.* 51.

So held, notwithstanding the improvement was intended to be a village improvement local in its character, from which the inhabitants of the town at large would derive little or no advantage, and the obligors, as residents of the village, or owning property there, were, for the sake of the benefit to the village, willing to assume the cost of the improvement, and indemnify the town. *Ibid.*

Commissioners of highways in exercising their authority in respect to the laying out of highways, have no right to make conditions with parties interested. They have no right to say, that if they shall order a highway to be laid out, individuals shall assume, or become bound to pay the expense. *Ibid.*

LAW OF THE ROAD.

1 R. S. Part 1, Chap. xx, Title 13.

§ 1. "Whenever any persons traveling with any carriages shall meet on any turnpike road or public highway in this state, the person so meeting shall seasonably turn their carriages to the right of the center of the road, so as to permit such carriages to pass without interference or interruption, under the penalty of five dollars for every neglect or offence, to be recovered by the party injured."

1 R. S. 695.

The true meaning of this is, that all persons thus meeting shall keep to the *right of the centre of the* WORKED *part of the road.* 7 *Wend.* 185. It is not the center of the *smooth* or *most traveled* part of the road which is the dividing line, but *the center of the worked part;* although the whole of the smooth or most traveled path may be upon one side of the center. It is no defence that the party had no design to offend, that he attempted to prevent a collision, that the road on his side was rough and rutty, and that it was more difficult for him than for the other party to turn out; unless the obstacles to turning out are insuperable or extremely difficult, he is without excuse. *Earing vs. Lansing,* (7 *Wend.* 185.)

There is no law of the road requiring a man on horseback, when meeting a horse or vehicle, to turn out on the *right* or *left* side. The rider must govern himself in this respect according to his notions of prudence at the time, under the circumstances. *Dudley vs. Bolles,* (24 *Wend.* 465.)

All the land within the highway fence is not necessarily subject to the right of way; and if not, it may be occupied by the owner; and if he place an obstruction there, and another is injured by it, he is not therefore liable; and if in the highway he is not liable unless the person injured exercised ordinary diligence to avoid it. 6 *Cow.* 189, *Cowen's Treaties,* 306 *and* 307.

The § 2 of this act provides for punishment of drivers addicted to drunkenness, and § 3 declares when they shall be discharged.

The § 4 provides a penalty against running horses on any public highway.

A penalty of $20, is inflicted on a driver of a conveyance for passengers, if he shall leave his horses untied, &c, while passengers are in the carriage.

All owners of carriages are made liable for the acts of drivers by the § 6, jointly and severally to the party injured; "and that whether the act occasioning such injury or damage be wilful or negligent. or otherwise, in the same manner as such driver would be liable."

A child of such tender age as not to possess sufficient discretion to avoid danger, is permitted by his parents to be in a public highway without any one to guard him, and is then run over by a traveler and injured, no action lies against the traveler, if there be no pretence that the injury was voluntary, or arose from culpable negligence on his part. *Hurlfield vs. Roper,* (21 *Wend.* 615.

In an action for such injury, if there be negligence on the part of the plaintiff, there cannot be a recovery, and although the child by reason of his tender age, be incapable of using that ordinary care which is required of a discreet and prudent person, the want of such care on the part of the parents or guardian of the child, furnishes the same answer to an action by the child as would its omission on the part of the plaintiff in an action by an adult. *Ibid.*

The same rule, it seems, would apply in an action by a blind or deaf man or a person non compos, who, under similar circumstances received an injury on a public highway. *Ibid.*

An action for negligence cannot be sustained, if the wrongful act of the plaintiff, co-operated with the misconduct of the defendant, to produce the damages sustained.

This is so whether the plaintiff's act be negligent or wilful.

Therefore, where domestic animals stray upon land where they have no right to be, and are injured by the negligence of the owner of the land, an action will not lie by the owner of the beasts against the owner of the land,

Otherwise, where the act by which the injury is effected is criminal, or is calculated to endanger human life.

If cattle driven along a highway, escape into an adjoining field, against the owner's will, the rule does not apply, and the trespass is excused.

And so of other cases of unavoidable necessity. Tonawanda railroad company vs. Munger, 5 Denio, 255.

If a man's servant, in the ordinary course of his business, obstruct the highway, from which a traveller receives a special injury, the master is liable. *Harlow vs. Humiston*, 6 *Cro.* 189.

The question seems to be, whether the act be such that he can justify himself to his master. If he may, it shall be beemed in the course of his business as a servant, and the master is liable, Ib.

Thus, where the master had been accustomed to lay his wood in a certain place for several years, and his servant laid the master's wood in the same place, while the latter was sick, and without his knowledge, *held*, that this was an act in course of his business as a servant. Ib.

An action for erecting a fence across a public highway, will not lie for special damage to an individual, by reason that the highway was more contiguous, and therefore more beneficial, and the deprivation of its use therefore more injurious to him than others. *Pierce vs. Dart.* 7 *Cow.* 609.

He must sustain some special damage to warrant an action. Ib.

It has been held that the action may be maintained for an injury received by being run against in the highway, even though done accidentally; the defendant being no otherwise to blame than driving in the wrong side of the road in the dark. 5, *T. R.* 648. 1 *Camp.* 497. 2 *East.* 593. 10 *Bing.* 112.

If the injury happen by *unavoidable accident*, no action will lie; otherwise, if blame attach to the defendant, though he be innocent of any intention to injure ; as if he drive a *too spirited horse*, or an *unbroken horse*, or pull a *wrong rein*, or use *imperfect harness.* 1 *Bing.* 213. 1 *Cow. Tr.* 3 *ed.* 245.

The party prosecuting must show himself free from negligence. A person being on the wrong side of the road, is *prima facia* evidence of negligence on his part. Where the streets are sufficiently wide, and a carriage is on the wrong side, it does not justify another in crossing over and interfering with it, merely because the law authorizes him to drive on that side ; and if a person do so, and an injury ensues, it is his own folly, and no action lies. 3 *Esp.* 685. *Ib.* 44. *Ib.* 278.

The rule of the road is not to be adhered to, if by departing from it an injury can be avoided. Yet in cases when parties meet on a sudden, and an injury results, the party on the wrong side should be held answerable, unless it appears clearly, that the party on the right had ample means and opportunity to prevent it. 3 *Car. & Payne* 554. 1 *Cowen's Tr. 3rd ed.* 346.

An action also lies against one who obstructs the highway, in consequence of which another is injured by running against the obstruction; but the plaintiff must in this, as in the other cases just mentioned, exercise ordinary care himself; and if there be an unlawful obstruction in the highway, plain to be seen at such distance as would enable the plaintiff to avoid it, and he nevertheless rides with careless violence against it, and is injured, he cannot sustain an action, for the injury arises from his own fault. He must use ordinary diligence to avoid it. 11 *East.* 60. 6 *Cow. R.* 189.

When a man in the employ of another, building a chimney, lays lime in the road, by which another's carriage is overturned and injured, the employer is liable; and also, even had the person employed done it by his servant. 1 *Bos. & Pull.* 404.

So the case of wood laid in the highway. 6 *Cow.* 189.

All the land within a highway fence, is not necessarily subject to the right of way; and if not, it may be occupied by the owner; and if he place an obstruction there, and another be injured by it he is not therefore liable.

OF WAYS.

A right of way is the privilege which one person or a particular description of persons, may have of going over another man's ground. It is a right pertaining to the realty—a right which one has in another's land, termed in law an incorporeal hereditament. *Nelson J.* 12, *Wend.* 98, 20 *Wend.* 99, *Conner vs. New,* 1 *Black* 45, 3 *Cruise* 72, 5 *Mass.* 129. 5 *Barn. & Cr.* 221.

In the case cited above of 20 *Wend.*, Cowen justice says, "If what appears to be a common highway, be not in fact laid out as such, or the public have refused to accept it as a donation, it is not a common road, and then the adjoining owner cannot claim it; for it is still the private property of the original proprietor, subject, perhaps, to a mere right of private way in the adjoining owner; provided he has taken a deed which bounds him upon it as a street."

This right of way is said to be only an *easement* upon the land of another. Hence trespass does not lie for its obstruction. On the same principle, if freehold, it must be created by deed; and it is entirely different from a public highway, and, in many respects, different from a private road. It may arise either,

1. By grant;
2. By prescription;
3. From necessity.

1. By grant: as when the owner of a piece of land grants to another the liberty of passing over his land in a particular direction; the

grantee thereby acquires a right of way over those lands. This right can be created by a reservation in a deed of land, (*White vs. Crawford,*) 10 *Mass. Rep.* 183; 14 *Mass.* 49. A. can purchase of B., by deed, a right to pass over his farm, or through a particular field, or to go to a spring, or creek, &c. The deed can be so limited in its terms, as that A. may have the right of passing or repassing when he pleases, or his family, or on horseback, or with carriages; or it may be limited to certain purposes, as to get water, go to mill, or go to church.

It is a principle of law that nothing passes as incident to the grant of an easement, but what is requisite to the enjoyment of the privilege. (*Lyman vs. Arnold,*) 5 *Mason Rep.* 195.

In giving his opinion in this case, Justice Story says, "In the construction of grants, that is doubtless to be adopted, which gives entire and liberal effect to the intention of the parties. When the object is distinctly seen, the ordinary means by which it is to be attained, are presumed to be within the purview of the parties. If the use of a thing is granted, whatever is necessary for the enjoyment of such use, or for the attainment of such use, is, by implication, granted also. But if it be not necessary, but may be a convenience only, it is not granted. So, too, grants are to be construed according to the subject matter, and the natural presumptions arising from their terms, and thus to render them expositions of rational intentions. The mere fact that a person having a grant of a privilege, servitude, or easement, in the land of another, bestows his labor upon the soil, or separates it, and gives it value thereby, constitutes no sufficient ground to infer a change of property in the soil ; for such labor is bestowed in order to enjoy such privilege, servitude, or easement." And the learned judge cites many authorities which fully establish these principles.

A right of way *ex vi termini,* imports a right of passing in a particular line, and not the right to vary it at pleasure, and go in different directions. This would be an injury to the owner of the land and an abuse of the right, (*Jones vs. Percival,*) 5 *Pick. Rep.* 485. It is held in this case, that "a custom for the inhabitants of a certain place, or the owners of a certain close, to pass over the soil of another, wherever it is most convenient to themselves, and least prejudicial to the owner, would be unreasonable. In practice, it would lead to contention and litigation. The rights of the respective parties under such a custom, are so undefined and doubtful, that the custom would be void for uncertainty."

But where a right of way is granted without any designation of the place in the deed, it becomes located by usage for a length of time; and being so located, it cannot be afterwards changed by the grantor. But if changed, the grantee, by usage of the newly designated way for a length of time, will be deemed to have acquiesced in the change. (*Wynkoop vs. Burger,*) 12 *John. Rep.* 222.

The grantee of a private way which has become foundrious and impassable, cannot, without being a trespasser, go on the adjoining close, and thus pass around the obstruction.

The rule is the same where the owner of the close through which the private way passes, caused the obstruction.‖

It seems that the only remedy for the owner of the way, is to remove the obstruction, and to prosecute for damages.

There is no distinction, with respect to the right of passing *extra viam*, between a private way by grant, and a private way by necessity, after the latter has once been selected or assigned.

The same rule applies to a private way by prescription, that controls in the case of a grant. (*Williams vs. Safford*,) 7 *Bar. S. C. Rep.* 309.

A way *in esse*, may pass by bargain and sale, but a way *de novo*, only by grant or lease. *Hays vs. Richardson*, 1 *Gill. & J.* 366.

Where a person granted a right of way over his land to another, the grantee at common law was bound to repair. And it is held in 12 *John. R.* 222, (cited above,) that "a private way being for the accommodation of the grantee, must be kept in repair by him at his own expense."

2. By prescription and immemorial usage: the true nature of this claim, is, that a person or certain number of persons have so long been accustomed to pass over the lands of another, that the law will presume that there was originally a grant. *Uninterrupted enjoyment* of a way for twenty years, without evidence that it had been used by leave, favor or mistake, is sufficient to leave to the jury to presume a grant. It must be an uninterrupted use of twenty successive years. *Odierne vs. Wade*, 5 *Pick.* 421. *Levitt vs. Wilson*, 3 *Bing.* 115. 3 *E.* 294. 22 *Pick.* 15. 7 *Met.* 33.

Uninterrupted use, *acquiescence* and *identity*, are said to be necessary for the presumption of a grant. 2 *Hilliard* 23, 5 *Conn.* 305, 3 *Mc Cord.* 131. The continued use should have been under a claim of right, or in other words, the use must have been adverse.

Hence, when a way is reserved by deed and the grantor and those claiming under him have used it in a mode nearly corresponding to the terms of the reservation, though long enough to give a prescriptive title, the use must be intended to have been under the deed, not adverse, and therefore the right will be limited by the terms of reservation. *Atkins vs. Boardman*, 2 *Met.* 457.

These prescriptive rights are *stricti juris*. A right of way for one purpose does not necessarily include a right of way for another purpose. The extent of the right must depend upon the circumstances *Ballard vs. Dyson*, 1 *Taunton* 279. *Conling vs. Higginson*, 4 *M. & W.* 245.

If it be a mere personal right, it cannot be assigned to any other person, nor transmitted by descent. *Finch's Law*, 17 31, *Year. B.* 7 *H.* 4, 36 *B.*

But when a right of way is annexed or appendent to an estate, it may pass by assignment when the land is sold. 6 *Mod. R.* 3, 2 *Lord Raym.* 922. (*Newmarch vs. Brandling*,) 3 *Swanston*, *C. R.* 99.

It has been held in this state, in the case of "*ex parte Coburn*," reported in 1 *Cowen R.* 568, "that a right of way is assignable property. It is a real or chattel interest, according to the term of its duration, and the former is well known in the law, as that sort of real property belonging to the class of incorporeal hereditaments. Not so of a license to enter on another man's land, without consideration. This is not an interest. It is a mere authority, revocable at any moment, not in its nature assignable, but limited to the person of the grantee. Giving permission to walk across another's land, is but an excuse for a trespass." The practice of passing over an unenclosed forest, common to all, is not sufficient to give a right by prescription.

So when the plaintiff's, and other neighboring occupants, passed over unenclosed land of the defendant, and the way was changed with the fields and fences at the will of the defendant, held, that the plaintiff's acquired no prescriptive right against the defendant. (*Trumbull vs. Rivers*) 3 *Mc Cord.* 131.

Every immaterial change will not destroy the identity of a way. But the question will depend on the situation of the country. 2 *Mc Cord.* 445.

Where one neglects to remove a house, wrongfully erected on his own land, and which obstructs a way belonging to him, this is no abandonment of the right of way. An erection of a house by himself might be an abandonment. (*Roger vs. Stewart,*) 5 *Vermont R.* 215.

A way cannot be extinguished by parol release. Such a release is a mere license which may be revoked. 5 *Har. & J.* 467.

Whether a way is public or private, is a question for the jury if the evidence is conflicting.

An obstruction placed in a private road, by the owner of the land over which it is laid, cannot be removed by one having no right to use the road. *Drake vs. Rogers,* 3 *Hill R.* 604.

The erection of a gate across a way claimed by the plaintiff, which is opened and shut at pleasure by all who pass, is not such an obstruction as would have the effect to extinguish the plaintiff's right, or of barring him of his remedy, however long it may have been erected. *Barnwell vs Magrath,* 1 *McMullan,* 174.

3. A right of way may arise from *necessity:* which is also founded on an *implied grant.*

If a man sells land to another which is wholly surrounded by his own land, in this case the purchaser is entitled to a right of way over the land of the vendor, to pass and repass to his own premises. This is founded on this principle of law, that whatever is necessary to the enjoyment of the grant, shall pass as incident to the grant; and without a passage to the land the grant would be useless.

This right of way by necessity exists only where the party claiming it has no other means of passing from the estate into the highway. 21 *Pick.* 278.

In this case the owner of the land over which the right of way exists, may assign or designate the way, and it will be sufficient if he give him convenient way of egress and ingress.

If the grantor decline to exercise such right, the grantee may select for himself, and will be supported in his selection, unless chargeable with palpable abuse. The grantee is bound to keep the way in repair, and is not permitted to go *extra viam,* out of the way, when the way is impassable, except, it seems, when the way is temporarily or accidentally obstructed. *Holmes vs Seely,* 19 *Wend.* 507.

A right of way from *necessity*, is to be construed strictly, and this right will cease with the necessity for it; as when the owner acquires the power of passing as directly over his own land. 2 *Bing.* 76 14 *Mass.* 55, 3 *Ib.* 492, 15 *Conn.* 39.

It was settled in the case of Clark *vs* Rugge, 2 Roll. Abr. 60, that if a man, having a close surrounded by his own land, grants such close to another, the grantee and those claiming under him, have a right of way of necessity through the lands of the grantor, as incident to the grant; which way of necessity the grantor may assign to the grantee through such part of his land as he shall think proper.

The same principle also seems to apply to the case when the close granted is partly surrounded by lands of a third person.

The case of *Henton vs Frearson*, 8 *Term. Rep.* 50, also appears to be a decision in favor of the principle that there is an implied right of way over the adjoining lands of the grantor, where the grantee has no other means of obtaining access to the lands granted, without trespassing upon the lands of a stranger.

The right of way of necessity over the lands of the grantor, in a conveyance in favor of the grantee, and those subsequently claiming the dominant tenement under him, is not a perpetual right of way; but it continues only so long as the necessity exists. If the grantee obtains another convenient way, the way of necessity ceases. Not so, when a person has a right of way by prescription or express grant. *N. Y. Life In. and Trust company vs Milnor*, 1 *Bar. C. R.* 354.

Sergeant Williams is of opinion that the right of way, when claimed by *necessity*, is founded entirely upon grant, and derives its force and origin from it. It is either created by express words, or it is created by operation of law, as incident to the grant; so that in both cases, the grant is the foundation to the title. *Note* 6 to 1 *Saun. Rep.* 323.

A distinction has been made between a way of necessity, and a way which is merely an easement or convenience, implying that the term easement is not applicable to that class of ways. 3 *Cruise* 76, 3 *Kent* 423.

Where a way is merely a matter of ease and convenience, the right to it does not arise by necessity, but by consent or agreement, and may be extinguished in like manner by unity of possession. 3 *Rawle*, 495.

But a right of way existing from necessity is not extinguished by the unity of possession, as a right of way to a church or market, or a right to a gutter carried through an adjoining tenement, or to a water course running over the adjoining lands. *Jordan vs. Atwood, Owen's Rep.* 121, *Cruise dig.* 23 24.

The right of way is an interest in lands, and, according to the statute of frauds, the right can only pass in writing.

Method of proceeding to obtain private way or road. Revised Statutes, 3rd Ed. p. 6. 32, § 93 94 95.

MANNER OF PLEADING A WAY.

In pleading a way by prescription or grant, no doubt the particular grounds of the title must be set forth, and as a way of necessity, is in truth nothing else but a way by grant, it must be pleaded in the same manner. And, if its origin cannot any longer be traced, must be claimed by grant or prescription, and pleaded as such, or (in some cases) as a non-existing grant. *Dutton vs. Taylor,* 2 *Lutw.* 148.

And if there once existed an unity of possession, some authors have supposed it must be claimed by way of grant. The better opinion now is, that a way of necessity cannot be pleaded in general terms. Indeed, it seems there is no general way of necessity, without specifying the manner whereby the land over which it is claimed, becomes charged with the burden.

Such was the decision in *Bullard vs. Harrison,* 4 *M. and Sel.* 387; *Boyce vs. Brown,* 7 *Bar. S. C. R.* 85.

In 6 Mod. 3, Staple *vs.* Hoydon, it is said, that if H. have a close surrounded by the land of another, A. for necessity, has a way over a convenient part of the land to his own close, or a necessary incident to his close. As if a self created necessity could be, either in law or reason, any justification of a tresspass committed on another's land.

It will be recollected that there are three sorts of private ways:

1. By grant.
2. Prescription.
3. Of necessity.

There seems to be no doubt, that whoever justifies under either of the two former titles, must set forth the particular ground of his title; as if it be by gran-, he must show it. Now a way of necessity, when it is considered, will be found to be nothing else but a way by grant. It derives its origin from a grant; and of course it is as necessary to set forth the title to a way of *necessity,* as it is to a way by grant.

Where the fact is, that there existed at one period an unity of possession, and there has been an uninterrupted use of the way for a long time, in oder to dispense with the necessity of pleading the grant with *profert,* which is generally required, it must be pleaded as a non-existing grant. Otherwise the case would be attended with insurmountable difficulty where the claimant is to set out his right in a plea to an an action of tresspass. 2 *Saun.* 175.

But where there has not been an unity of possession, and the way has been used immemorially, it must then be claimed as a way by prescription. 3 *Term Rep.* 157.

Therefore a way of *necessity* is not extinguished by unity of possession. For unity of possession appears to be the foundation of the right.

It must be stated, that the same person was seized in fee of both closes, *simul et semel*, and being so seized, he granted one of them. *Pomfret vs. Ricroft*, 1 *Saun. Rep.* 323.

Under a plea of right of way, in an action of trespass, *quare clausam fregit*, the defendant is at liberty to prove a right of way in himself, either by producing deeds to show it, or by parol evidence of twenty years' uninterrupted use, adverse, or in hostility to, the the owner of the land, from which a grant may be inferred.

And a defendant after having undertaken to an prove express grant or reservation of a way, by deed, may prove one by prescription. He is not bound to elect his mode of proof, and to abide by such election. *Hamilton* vs. *White*, 4 *Bar. S. C. R.* 60.

PLANK ROAD LAW.

LAWS OF NEW YORK, 1847.

AN ACT *to provide for the incorporation of companies to construct plank roads, and of companies to construct turnpike roads. Passed May* 7, 1847.

The People of the State of New York, represented in Senate and Assembly, do enact as follows:

§ 1. Any number of persons not less than five, may be formed into a corporation, for the purpose of constructing and owning a plank road, or a turnpike road, by complying with the following requirements: Notice shall be given in at least one newspaper, printed in each county through which said road is intended to be constructed, of the time and place or places where books for subscribing to the stock of such road will be opened; and when stock to the amount of at least five hundred dollars for every mile of the road so intended to be built, shall be in good faith subscribed, and five per cent paid thereon, as hereinafter required, then the said subscribers may, upon due and proper notice, elect directors for said company; and thereupon, they shall severally subscribe articles of association, in which shall be set forth the name of the company, the number of years that the same is to continue, which shall not exceed thirty years from the date of said articles, whether it is a plank road or a turnpike, which the company is formed to construct; the amount of the capital stock of the company; the number of shares of which the said stock shall consist; the number of directors, and their names, who shall manage the concerns of the company for the first year, and shall hold their offices until others are elected; the place from and to which the proposed road is to be constructed; and each town, city, and village into or through which it is intended to pass, and its length as near as may be. Each subscriber to such articles of association, shall subscribe thereto his place of residence, and the number of shares of stock taken by him in said company. The said articles of association, may, on complying with the provisions of the next section, be filed in the office of the secretary of state, and thereupon, the persons who have so subscribed, and all persons who shall from time to time, become stockholders in such company, shall be a body corporate, by the name specified in such articles, and shall possess the powers and privileges, and be subject to the provisions contained in titles three and four, chapter eighteen, of the first part of the Revised Statutes.

It is no objection to the validity of a subscription to the capital stock of a plank road company, that it was made upon a separate paper, which only a portion of the stockholders had subscribed; there having been several similar papers used in lieu of the *books* required by the act to be opened in different places for subscription. *Hamilton & D. Pl. R. co. vs. Rice,* 7 *Barb.* 156.

Nor is it any objection to the validity of such a subscription, or the right of the company subsequently organized, to maintain an action upon it, that at the time it was made, there was no company in existence. *Ibid.*

All the cases which have held the subscriptions of stock void, are based either upon the supposed want of necessary parties to the agreement; of a sufficient consideration to uphold it; or of a sufficient promise expressed in it.

And where a subscriber agreed to become a member of a proposed company, as soon as the amount of stock required by this law should be subscribed, and to pay the amount of his subscription when the company should be organized; it was held by the court, that as soon as stock to the amount of $500 for every mile of the road was subscribed, and five per cent paid in, and the company organized, he was liable to be sued by the company for the installments due upon the subscription. *Ibid.*

§ 2. Such articles of association shall not be filed in the office of the secretary of state, until five per cent on the amount of the stock subscribed thereto, shall have been actually and in good faith paid, in cash to the directors named in such articles, nor until there is endorsed thereon, or annexed thereto, an affidavit made by at least three of the directors named in such articles, that the amount of capital stock required by the first section has been subscribed, and that five per cent on the amount has actually been paid in.

§ 3. A copy of any articles of association filed in pursuance of this act with a copy of the affidavit aforesaid endorsed thereon or annexed thereto, and certified to be a copy by the secretary of this state or his deputy, shall in all courts and places be presumptive evidence of the incorporation of such company, and of the facts therein stated.

§ 4. Whenever any such company shall be desirous to construct a plank road or turnpike road through any part of any county, it shall make application to the board of supervisors of such county at any meeting thereof legally held, for authority to lay out and construct such road, and to take the real estate necessary for such purpose; and the application shall set forth the route and character of the proposed road as the same shall have been described in the articles of association filed as aforesaid. Public notice of the application shall be given by the company previous to presenting the same to such board by publishing such notice once in each week for six successive weeks in all the public newspapers printed in such county, or in three of such newspapers if more than three are published in such county, which notice shall specify the time when such application will be presented to such board, the character of the proposed road, and each town, city, and village in or through which it is proposed to construct the same.

§ 5. If such company shall desire a special meeting of the board of supervisors for hearing the same, any three members of such board may fix the time of such meeting, and a notice thereof shall be served on each of the other supervisors of the county, by delivering the same to him personally or by leaving it at his place of residence at least twenty days before the day appointed for such meeting. The expenses of such special meeting and of notifying the members of such board thereof, shall be paid by such company.

§ 6. Upon the hearing of the said application, all persons residing in such county or owning real estate in any of the towns through which it is proposed to construct such road, may appear and be heard in respect thereto. Such board may take testimony in respect to such application, or may authorize it to be taken by any judicial officer of such county, and it may adjourn the hearing from time to time.

§ 7. If after hearing such application such board shall be of the opinion that the public interests will be promoted by the construction of such road on the proposed route as shall be described in the application, it may, if a majority of all the members elected to such board shall assent thereto, by an order to be entered in its minutes, authorize such company to construct such a road upon the route specified in the application, and to take the real estate necessary to be used for that purpose; a copy of which order certified by the clerk of such board the said company shall cause to be recorded in the clerk's office of such county, before it shall proceed to do any act by virtue thereof.

§ 8. Whenever any such board shall grant such an application, it shall appoint three disinterested persons who are not the owners of real estate in any town through which such road shall be proposed to be constructed, or in any town adjoining such town, commissioners to lay out such road; the said commissioners after taking the oath prescribed by the constitution shall proceed without unnecessary delay to lay out the route of such road in such manner as in their opinion will best promote the public interest; they shall hear all persons interested who shall apply to them to be heard, they may take testimony in relation thereto, they shall cause an accurate survey and description to be made of such route and of the land necessary to be taken by such company for the construction of such road and the necessary buildings and gates, they shall subscribe such survey and acknowledge its execution as the execution of deeds is required to be acknowledged, in order that they may be recorded, and they shall cause such survey to be recorded in the clerk's office of such county. If such company shall intend to construct its road continuously in or through more than one county, such application shall specify the number of commissioners which the company desire to have appointed to lay out such road, which shall not exceed three for each county, and an equal number of such commissioners shall be appointed by the board of supervisors of each county in or through which it shall be proposed to construct such road, but the whole number of such commissioners shall not be less than three, nor without the consent of such company shall it exceed six, unless the number of counties in or through which it is proposed to construct such road shall exceed that number. And the commissioners so appointed shall lay out the whole of such road, and shall make out a separate survey of so much thereof as lies in each county, which shall be subscribed and acknowledged as aforesaid and recorded in the county clerk's office of such county. Such company shall pay each of the said commissioners two dollars for every day spent by him in the performance of his duties as such commissioner, and his necessary expenses.

§ 9. No such road shall be laid out through any orchard to the injury or destruction of fruit trees, or through any garden, without the consent of the owner thereof, if such orchard be of the growth of four years or more, or if such garden has been cultivated four years or more before the laying out of such road, nor shall any such road be laid out through any dwelling house or buildings connected therewith, or any yards or enclosures necessary for the use and enjoyment of such dwelling without the consent of the owner, nor shall any such company bridge any stream where the same is navigable by vessels or steamboats, or in any manner that will prevent or endanger the passage of any raft of twenty-five feet in width.

§ 10. No plank road shall be made on the roadway of any turnpike company without the consent of such company, and any plank road company formed under this act shall have power to contract with any turnpike company for the purchase of the roadway or part of the roadway of such turnpike company on such terms as may be mutually agreed on; whenever a plank road shall be made as provided in this act on or adjoining the route of any turnpike road, the company owning such turnpike road is authorized to abandon that portion of their road on or adjoining the route of which a plank road is actually constructed and used; but nothing herein contained shall be so construed as to prevent any plank road from crossing any turnpike road, nor any turnpike road from crossing any plank road.

§ 11. The route so laid out and surveyed by the said commissioners shall be the route of such road, and such company may enter upon, take and hold, subject to the provisions of this act, all such lands as the said survey shall describe as being necessary for the construction of such road and the necessary buildings and gates. But before entering upon any of such lands, the company shall purchase the same of the owners thereof, or shall, pursuant to the provisions of this act, acquire the right to enter upon, take and hold the same.

§ 12. If any owner of such land shall from any cause be incapable of selling the same, or if such company cannot agree with him for the purchase thereof, or if, after diligent inquiry the name or residence of any such owner cannot be ascertained, the company may present to the first judge or county judge of the county in which the lands of such owner lie, a petition setting forth the grounds of the application, a description of the lands in question and the name of the owner if known and the means that have been taken to ascertain the name and residence of such owner, if his name and residence has not been ascertained, and praying that the compensation and damages of the owner of the lands described in the petition may be ascertained by a jury. Such petition shall be verified by the oaths of at least two directors of the company, and if it shall alledge that the name or residence of any owner is unknown, it shall be accompanied by affidavits proving to the satisfaction of the said judge that all reasonable efforts have been made by the company to ascertain the name and residence of any owner whose name or residence is unknown.

§ 13. On receiving such petition, the said judge shall appoint a time for drawing such jury which shall be drawn from the grand jury box of the county by the clerk thereof, at his office. At least fourteen days' notice of the time and place of such drawing shall be served personally upon each owner of lands described in the petition, who shall be known and reside in the county where the lands lie or by leaving the same at his residence, and such notice shall be served on all other owners in the manner aforesaid or by putting the same into the post office directed to them at their respective places of residence and paying the postage thereon, or by publishing the same once in each week for two successive weeks in a newspaper printed in such county, the first of which publications shall be at least fourteen days before such drawing.

§ 14. In case any lands described in such petition shall be owned by a married woman, infant, idiot or insane person, or by a non-resident of the state, the said judge shall appoint some competent and suitable person having no interest adverse to such owner to take care of the interests of such owner in respect to the proceedings to ascertain such compensation and damages. And all such notices as are required to be served on any owner residing in such county, shall be served upon the person so appointed in like manner as on such owner; but any person so appointed to take care of the interests of any such non-resident may be superseded by him.

§ 15. The said judge shall attend such drawing and shall decide upon any challenge made to any juror drawn by any person interested. Twenty-four competent and disinterested jurors and as many more as the said judge shall direct shall be drawn; the clerk shall make, certify and deliver to the judge and to any party requiring the same a list of them, and the ballots drawn shall be returned to the box. The said judge if he shall deem it necessary, may at any subsequent time direct the drawing of an additional number of jurors, and they shall be drawn, and all such proceedings in relation to such drawing shall be had in the manner hereinbefore provided. Before proceeding to draw any such jury the company shall furnish to the said judge proof by affidavit satisfactory to him, of the time and manner of serving and publishing notice of such drawing, which affidavit shall be filed in such clerk's office; and no jury shall be drawn unless it shall appear to the satisfaction of the said judge that the provisions of this act in respect to giving notice of such drawing have been complied with.

§ 16. From the jurors so drawn the said judge shall draw as many as he shall deem necessary to secure the attendance of twelve, and he shall issue his precept directed to the sheriff of such county, either of his deputies or any constable of such county to summon the jurors so drawn by the said judge, to attend at the time and place therein specified, to ascertain such compensation and damages. And he may from time to time, in case of the absence or inability to serve of any juror directed to be summoned, draw and direct to be summoned as aforesaid, as many as may be necessary in his opinion to secure the attendance of twelve.

§ 17. Every juror named in any such precept, shall at least four days before the day therein specified for his attendance, be summoned personally, or by leaving at his residence, a notice containing the substance of such precept. The officer serving such precept, shall return it to the said judge, with an affidavit of the manner of serving the same, and of the distance necessarily traveled by him for that purpose; and such officer shall receive for making such service, six cents a mile for the distance so traveled.

§ 18. Every juror so summoned, who shall neglect or refuse to attend or serve, in pursuance of such summons, shall be liable to the same penalties, as in case of such neglect or refusal of a person duly summoned as a juror in a court of record, and may be excused by the said judge from attending or serving, for reasons for which such juror might be so excused if summoned as a juror in such court. Every juror attending, shall be entitled therefor to one dollar a day, and his reasonable and necessary expenses to be paid by the company.

§ 19. On the application of any party interested, any judge or justice of the peace, may issue a subpœna requiring witnesses to attend before such jury, and such subpœna shall have the same force and effect; and witnesses duly subpœnaed by virtue thereof, and refusing or neglecting to obey the same, shall be subject to the same penalties and liabilities as though the subpœna were issued from a court of record, in a suit pending therein.

§ 20. The time and place of meeting of the jury, to ascertain such compensation and damages, may be fixed by the said judge, by an order to be made by him at any time after receiving such petition; and notice thereof shall be served on the owners whose lands are described in the petition, as follows: on any owner residing in the county, or within fifteen miles of the lands in question owned by him, personally, or by leaving the same at his residence, at least fourteen days before the time so fixed; on any other owner residing within this state, and whose residence is known, in the manner aforesaid, or by putting the notice into the post office, directed to him at his place of residence, and paying the postage thereon: on any owner residing out of the state, and not within fifteen miles of the lands in question, owned by him, by putting the notice in the post office, directed and paid as aforesaid, at least forty days before the time so fixed; and on owners whose residence is unknown, by publishing the notice once in each week, for six successive weeks, in one of the public newspapers printed in the county.

§ 21. The jurors so summoned, shall meet at the time and place fixed by the said judge for that purpose, and shall be sworn by him to diligently inquire and ascertain the compensation and damages which ought justly to be paid for the lands described in the petition, or for those of them in respect to which they shall be called upon to enquire, to the owners thereof, and for taking the same for such road, and faithfully to perform their duty as such jurors, according to law.

§ 22. The said judge shall attend such jurors, shall administer oaths to witnesses called before them, shall take minutes of the testimony given, and admissions of the parties made before them, shall advise such jury as to the law applicable to any case that may arise, shall receive, certify and return to the county clerk's office, the verdicts agreed upon by them, and while so attending, shall have all the powers possessed by a court of record, when trying issues of fact joined in civil cases.

§ 23. The jury after hearing the parties, and viewing the lands in question, in each case, shall, by a verdict, ascertain and determine the compensation and damages that ought to be paid to the owner for the land, to be taken by the company, and for taking the same for such road, and also the amount that ought to be paid to him for the time spent, and necessary expenses incurred by him in respect to the proceedings, to ascertain and determine such compensation and damages, of which time and expenses a bill of items shall be presented to the jury, verified by the oath of the owner or his agent, and such compensation and damages shall be ascertained and determined without any deduction on account of any real or supposed benefit, which the owners of such lands may derive from the construction of such road.

§ 24. Such jury shall not proceed to a hearing in any case until the company shall have produced to the said judge, satisfactory proof by affidavit, that the notice of the jury has been given in such case, according to the provisions of this act; and such affidavit shall be attached to and filed with the certificate of the verdict in the case. And on any such hearing, no evidence or information shall be given, nor any statement made to the jury, of any proposition by, or negotiation between the parties or their agents, in respect to any such lands, or such compensation or damages, nor shall any such petition contain any such statement or information.

§ 25. Such jury, finding any such verdict, shall, after agreeing upon the same, make a certificate thereof, and sign and deliver the same to the said judge; and shall embrace therein a particular description of the land, in respect to which it is found. Such certificate may include one or more verdicts, in the discretion of the jury. Every such certificate shall be certified by the judge to have been made by such jury and shall be recorded in the records of deeds in the clerk's office of the county where the lands therein described shall lie, at the expense of the company.

§ 26. Whenever it shall become necessary for any such company to use any part of a public highway for the construction of a plank or turnpike road, the supervisors and commissioners of highways of the town in which such highway is situated, or a majority, if there be more than one such commissioner in such town, may agree with such company upon the compensation and damages to be paid by said company, for taking and using such highway for the purposes aforesaid. Such agreement shall be in writing, and shall be filed and recorded in the town clerk's office of such town. In case such agreement cannot be made, the compensation and damages for taking such highway for such purpose, shall be ascertained in the same manner as the compensation and damages for taking the property of individuals. Such compensation and damages shall be paid to the said commissioners, to be expended by them in improving the highways of such town.

The constitutionality of this law has been doubted by many. As the owner of the adjacent lands owns to the centre, and has the whole fee, except the *easement,* of the highway, how can a plank road company occupy it, and place their materials in it, and erect gates, without compensation to the owner ?

The only published decision on that question is here inserted. The Court of Appeals have never passed any rule on the question.

SUPREME COURT—JEFFERSON GENERAL TERM.

Benedict vs. Goit. Kirkland for the Plaintiff : Comstock for the Defendant. Gridley, Justice.

The questions raised upon the demurrer in this cause, involve the constitutionality of one of the provisions of the act organizing the "Rome and Oswego Road Company," and authorizing the construction of the road known as the plank road, leading from Rome to Oswego. By the second section of the act (laws of 1844, p. 434,) the provisions of the general statute concerning turnpike corporations, (1 R. S. 581,) are adopted and made applicable to the road in question. The 29th section of the general act confers the power of taking any public highway, by appraising the value of the public interest in the road, and paying the same to the commissioners of highways, to be by them applied to the improvement of the roads in their respective towns. It is this particular provision to which the plaintiff objects. He lives upon the line of the road which has been constructed upon the site of a public highway that has been appropriated under this section; and he complains that the defendant has committed an unwarrantable trespass upon his land by the act of building the road—and also that in grading it an embankment has been raised in the front of his dwelling house, so that he has been compelled to incur considerable expense and inconvenience in raising his buildings to the level of the road. The defendant has in his plea justified the alleged trespass, under the act, as a servant of the company, and to this plea the plaintiff has demurred.

The ground assumed by the plaintiff is that the "*locus in quo,*" since its appropriation by the company, has ceased to be a public highway, and therefore that the entire interest in it has reverted to him as the original owner; and doubtless the conclusion follows as a legal consequence if the premises are admitted to be correct. Several cases have been decided in this state, holding that railroad companies, which have occupied portions of the public highway, are liable to the owners of the soil for an unlawful invasion of their rights. (3 Hill, 567; 25 Wendell, 462; 5 Hill, 170.) But a railroad is in no sense a public highway. The nature of the road forbids its use by the public in common with the company. But the position assumed by the defendant's counsel is, that the road in question in this suit is a *public highway still,* open for the public use, precisely as is every public road in the state ; and that the corporation organized by the act before cited, has succeeded to all the rights and powers of the commissioners of highways, in the several towns through which the road passes. We are of this opinion.

1. That the road is public in the sense that every citizen has the right to travel on it, either on foot, on horseback, in his carriage, or with his team, subject to the payment of the legal tolls, is undeniable. The object of the legislature in passing the act, seems to have been two-fold: 1st, to provide for a better road than is ordinarily constructed under the supervision of the town commissioners. In order to carry out this design, the sixth section of the act declares that the "track of said road shall be constructed of timber, plank, gravel, or other hard material, so that the same shall form a hard, smooth and even surface." It cannot be doubted that the legislature has the power to ordain by law that all the public highways in the state shall be graded, leveled, and constructed of some hard substance, under the direction of the town commissioners, so as to form a hard, smooth and even surface. And it is equally clear that the commissioners in the respective towns, by virtue of the powers they now possess, may in their discretion improve the public highways in all these respects, to the extent of the means under their control. No importance should be attached to the idea of a plank road; for the intent of the act was simply to provide for a road constructed of some hard and durable material. A gravel road would equally have satisfied the requirements of the act.

2d. A further object of the legislature was to provide for a different mode of keeping the road in repair. It devolves the *duty* on the incorporation instead of the town commissioners, and provides for raising the *necessary funds* by levying tolls on those who travel on the road, instead of resorting to an assessment upon the inhabitants of the respective districts. We do not see why these provisions are not entirely within the power of the legislature. May they not vest the oversight of the public roads in any other agent than the commissioners? And may they not provide any other mode that seems to them more equitable and expedient for keeping the roads in repair, than the ordinary highway assessment? And if these provisions should be extended to all the public highways in the state, would they cease to be *public highways ?* And would the public interest in the roads become forfeited by such an act, and the owners of the fee of the lands over which the roads are constructed, succeed to their original estate in those lands, discharged of the right of the public to use them as public highways ? We think not. And if this consequence would not follow from a general exercise of this power by the legislature, we do not see why it should in the particular instance in question. To deny the power of the legislature to exercise this extensive control over the public highways in the state, would be to interpose an effectual barrier to any important improvement in our public roads and thoroughfares, and would operate most injuriously upon the advancement and prosperity of the commercial, manufacturing, agricultural and social interests of the community. We are of opinion, therefore, that, upon principle, the road which was taken by this company by virtue of the provisions of the act, did not cease to be a public highway; and that when the corporation paid the commissioners of highways for the public interest in the road, it succeeded to all the rights of the town commissioners to make such repairs in the road as the public interest required, whether such repairs consisted in excavations or embankments, to bring the road to a proper grade, and thus to improve its condition as a public thoroughfare.

II. This question seems to have been settled, so far as such a question can be settled by legislative authority. It is not pretended by the counsel for the plaintiff, that any difference exists between the rights acquired by this corporation, and those to which a turnpike company would be entitled under the like circumstances. The provision in question has been incorporated in the acts relating to turnpike companies for more than forty years. The original turnpike act was passed in 1807, and was re-enacted in the revision of 1813, and again in 1830; and the principle was again distinctly asserted in the 26th section of the act concerning plank and turnpike roads, passed in 1847. (Laws of 1847, ch. 210, p. 216, sec. 26.) During all this period, within which the state has been traversed in every direction by turnpike roads, in the construction of which the right which is now drawn in question must have been most extensively exercised—we have no knowledge that any individual has complained of the exercise of this right, or controverted the constitutionality of the act which conferred it. This long practice under the law, in addition to its repeated re-enactment, though not conclusive, nevertheless is strong evidence in favor of the correctness of our views upon the question under consideration. This however is not all. The point has been directly adjudicated in the case of the Commonwealth vs. Wilkinson, (16 Pickering 175.) In this case a turnpike road was adjudged to be a public highway.

It is not to be denied that the corporation have an interest, of a certain description, in the road. They may doubtless recover damages of any person who shall commit an injury upon it. This is, however, precisely the same interest which a public agent would possess in it, who should be employed under a legislative enactment to keep the road in repair, for a given compensation. Again, every citizen over whose lands a public road is laid out, retains an interest in the road. He still owns the fee of the land, and may maintain ejectment or trespass against third persons who take possession of it, or use it for any other purpose than a highway. The public interest in the road is merely that of a right of way; and we cannot see how the interest which any individual or corporation may have in the road, that does not intetfere with this right, can in any degree affect its character as a public highway. The interest which this corporation has in the road in question, is entirely compatible with the use of it as a public highway, subject only to the single condition of paying tolls; and this, as we have already seen, is a burden imposed expressly for the keeping of the road in repair. And in many instances it is far more equitable than the common mode by assessment of a highway tax. By the latter mode, a citizen whose employment consists in drawing heavy loads over a given tract of highway, in no degree indemnifies the public for the repairs which he renders necessary, by the insignificant assessment which he pays in his own district. In all such cases ths system of tolls is a far more equal and just one.

III. The claim made under the 4th count of the declaration must depend upon the decision of the question we have just considered. If the corporation has succeeded to the rights and powers of the commissioners of highways, then any inconvenience or damage which the plaintiff has suffered by proper and reasonable repairs of the public highway is "*damnum absque injuria.*" If the commissioners see fit, for the purpose of grading a highway, to cut down a hill, or to raise an embankment in the road adjacent to the premises and dwellings of citizens, by which those citizens suffer expense and inconvenience, no action can be maintained for the injury. The question was thus settled in the case of Graves vs. Otis, (2 Hill, 466,) where the injury complained of consisted in cutting down an eminence in a public street and sidewalk in the village of Watertown, by which the plaintiff's store was left some six or eight feet above the level of the sidewalk adjacent to the premises. (See also to the same point, 1 Pickering 418; 4 Term Rep. 794; 8 Cowen 146; Sedgwick on damages 110.

We have not inquired whether it was necessary to allege in the plea, that the value of the public interest in the road had been appraised and paid to the town commissioners, nor whether any other criticism may be applied to the plea. But we have assumed that the acts for which the action is brought, were done in the legitimate exercise of the powers conferred by the statute, and were incident to the right of constructing a road of a proper grade, and so as to form a hard, smooth and even surface, as directed by the 6th section of the act. In doing this, we have complied with the request of the counsel of both parties, their object being to obtain a decision of the important constitutional question which we have discussed. But for any unreasonable exercise of the power conferred by the act, the agent is responsible ; and if such a case can be made out, the plaintiff should have the opportunity to amend his declaration.

The demurrer is therefore overruled, with the right to amend on payment of costs.

§ 27. Any party interested in any such verdict may, within twenty days after being notified of the rendition thereof, apply to the supreme court for a new trial, and it may be granted upon such terms as to the costs of the application and of the first trial, as that court shall deem reasonable. If a new trial shall be granted, a jury shall be drawn therefor, and the same proceedings shall be had as are hereinbefore provided.

§ 28. Within forty days after the rendition of any such verdict, if a new trial shall not be applied for, the company shall pay to the person entitled to receive the same, the amount thereof, or shall make a legal tender thereof to him, if he shall refuse to receive the same; and the company may thereupon enter upon the lands in respect to which such verdict was rendered, and take and hold the same to it and its assigns, so long as it shall be used for the purposes of such a road as such company was formed to construct.

§ 29. If any person entitled to receive the amount of any such verdict be not a resident of this state, or cannot be found therein after diligent search, the company may furnish to the said judge satisfactory proof, by affidavit, of such fact, and he shall thereupon make an order that the amount of such verdict be paid to the treasurer of the county in which the lands lie, in respect to which such verdict was found, for the use of such owner, and that notice of such payment shall be given by publishing the same once in each week, for six successive weeks, in a newspaper published in the county. On satisfactory proof being made to the said judge, by affidavit, within three months from the time of making the last mentioned order, of such payment and publication, he shall make an order authorizing the company to take and hold the land in respect to which such verdict was rendered, in the same manner and with the same effect as if such payment had been made to the owner personally. The affidavit and orders mentioned in this section, and all other affidavits and orders made, and precepts issued in the course of the proceedings under this act, in relation to the acquisition of the land to be used for such road, shall be filed in the county clerk's office, and all such orders shall be recorded by such clerk in the records of deeds, at the expense of the company.

§ 30. If any owner shall apply for a new trial, the company, upon depositing the amount of the verdict sought to be set aside, in such manner as the said judge shall, upon hearing the parties, direct, in trust that the same, or so much thereof as the said owner shall be entitled to receive, shall be paid to him on demand, and on giving such security, by bond, as the judge shall approve, for the payment to such owner of any sum which he may be entitled to receive from the company, in respect to the land in question, by reason of any verdict or the judgment of any court, for such compensation, damages, costs and expenses, the company may enter upon and use such lands for the purposes of such road, but the title of the owner thereof shall not be divested until the payment or legal tender to him of the whole amount which he shall be entitled to receive from the company for such compensation, damages, costs and expenses; and on such payment or tender being made, the company shall be entitled to take and to hold such lands to it and to its assigns, so long as the same shall be used for the purposes of such road as such company was formed to construct.

§ 31. Every plank road made by virtue of this act, shall be laid out at least four rods wide, and shall be so constructed as to make, secure and maintain, a smooth and permanent road, the track of which shall be made of timber, plank, or other hard material, so that the same shall form a hard and even surface, and be so constructed as to permit carriages and other vehicles conveniently and easily to pass each other, and also, so as to permit all carriages to pass on and off where such road is intersected by other roads.

§ 32. Every turnpike road that shall be constructed by virtue of this act, shall be laid out at least four rods wide; and shall be bedded with stone, gravel, or such other material, as may be found on the line thereof, and faced with broken stone or gravel, so as to form a hard and even surface, with good and sufficient ditches on each side, wherever the same is practicable. The arch or bed of such road shall be at least eighteen feet wide, and shall be so constructed as to permit carriages and other vehicles conveniently to pass each other, and to pass on and off such turnpike where it may be intersected by other roads.

A turnpike road company is liable to an indictment at common law for suffering their road to be out of repair, &c. *President, &c. of the Sus. and Bath turnpike road company, vs. the People,* 15 *Wend.* 267.

When the road is out of repair, *prima facie,* all the directors are liable; those however, who have done their duty, may show facts to exonerate themselves, &c. *Kane vs. the People,* 8 *Wend.* 203.

A turnpike company are bound to erect and keep in repair bridges across streams, &c.

If bridge carried away, company must rebuild it within a reasonable time.

If road is impassible by sudden flood or other irresistible cause, the defendants in an action must, if they have an excuse, rejoin setting forth the cause of the injury; and then the jury will pass on the facts, &c. *The People vs. Hillsdale and Chatham turnpike road.* 23 *Wend.* 254.

§ 33. In each county of this state, in which there shall be any plank road, or turnpike road, constructed by virtue of this act, there shall be three inspectors of such roads, who shall not be interested in any plank or turnpike road in such county. They shall be appointed by the board of supervisors of the county, and shall hold their offices during the pleasure of such board. Before entering on their duties, such inspectors shall take and subscribe the constitutional oath of office, and file the same in the office of the clerk of the county.

§ 34. Whenever any such company shall have completed their road, or any five consecutive miles thereof, it may apply to any two of the inspectors to be appointed pursuant to this act, in the county where said road, or a part thereof, so completed and to be inspected, is located, to inspect the same; or, if such inspectors, or a majority of them are satisfied on inspection, that the road so inspected is made and completed according to the true intent and meaning of this act, they shall grant a certificate to that effect, which shall be filed in the offce of the county clerk.

The inspectors shall be allowed two dollars per day for their services, pursuant to this section, to be paid by the company whose road they inspect.

§ 35. Upon filing as aforesaid such certificate, the company owning any plank road so inspected, may erect one or more toll gates upon their road, but not within three miles of each other, and may demand and receive toll, not exceeding one and a half cents per mile, for any vehicle drawn by two animals, and for any vehicle drawn by more than two animals, one half cent per mile for every additional animal; for every vehicle drawn by one animal, three-quarters of a cent per mile; for every score of sheep or swine, and for every score of neat cattle, one cent per mile; for every horse and rider, or led horse, half a cent per mile. In no case shall any plank road company charge or receive rates of toll which will enable said company to divide more, nor shall any company divide more than ten per cent per annum on their capital stock actually paid in and invested in their road, after keeping the road in repair, and appropriating not exceeding ten per cent. per annum on their capital stock invested as aforesaid, as a fund for the reconstruction of their road when necessary.

The supreme court have decided in effect that the Co. are not bound to put up gates as often as every three miles; and if they do not they have a right to charge a person travelling the whole distance between the gates. The consequence would be if the Co. have but one gate, they could charge the traveller passing it the whole distance of the road.

In regard to this question of rates of toll, Judge Gridley has made the following decision. "By the 35th section of the general plank road act (Laws of 1847, P. 226) the Northern plank road company was authorized to erect one or more toll gates upon their road, but not within three miles of each other, and to demand and receive toll, not exceeding one and a half cents per mile, for any vehicle drawn by two animals, &c.

In the case of *Stewart vs Rich*, (1 Caines, 182,) a construction was given to a clause contained in the eleventh section of the Cherry Valley Turnpike act, by which the company was authorized to receive the tolls and duties in the act mentioned, and no more, that is to say, for any number of miles not less than ten in length of said road, the following sums of money, and so in proportion for any greater or lesser distance, viz. &c.

Recoveries had been had against a toll gatherer, for receiving full toll of persons who had not travelled ten miles on the road. The question before the court was, whether a deduction should not have been made, from the full toll, in proportion to the distance which the traveller had used the road. It was held that no such deduction need be made, but that full toll might lawfully be collected at every gate, irrespective of the distance which the traveller had used the road.—This construction says Kent, Justice, in delivering the opinion of the court, 'is the only one that is reasonable, and it will satisfy the words. The idea that the company must vary the toll at any ten mile gate, on the suggestion that a person has used the road for a less distance than ten miles, is inadmissible because impracticable. The toll gatherer has no means of knowing whether the traveller had rode ten miles or a less distance, previous to his arrival at the gate. If this suggestion was allowed to be a ground to reduction of toll, it would open a door to the greatest imposition and fraud upon the company. Again, in *The People* vs. *The Kingston and Middletown Turnpike Co.* (23 Wend 193) the same doctrine is reiterated, and the principle of the case in Caines reaffirmed The prevailing opinion of the court was delivered by Ch. J. Wilson. His language is very explicit. He says, 'it was also said that the demurer to the five last replications was mainly intended to raise and settle the construction of the act whether the rate of toll shall be in proportion to the *distance actually travelled*, or shall be determined by the *distance between the gates*, as located. The latter I am of opinion is clearly the rule intended by the act. (Laws of 1831, P. 49, §5,) By the section referred to, the company may erect the gates at such places as they see fit; but they can demand only, '*the following rates of toll for every ten miles, and in the same proportion for a shorter distance*,' &c.—This clause refers to the distance between the gates.' If that be five miles, the company may demand half toll. If it be two and a half miles, they may demand a quarter toll, and so in proportion. Judge Cowen dissented from a majority of the Court in this case, in the general result, but on this particular point he agreed with his brethren. Speaking of the construction claimed by the appellant's counsel as the true one, he says, 'such a construction would leave every traveller to estimate his own toll, and make it utterly impracticable for the toll gatherer to perform his duty. It would lay him open to constant imposition.' No rule, practicable in its application, can be laid down, by which any traveller can be charged with the exact amount of toll which would be due from him in proportion to the distance actually travelled. In some cases, where the traveller and the place of his residence are known to the toll gatherer, he may judge with a good degree of probable accuracy, how far the individual has travelled on the road; but in a great majority of cases it would be a mere matter of conjecture; and in no case would he know how far the traveller was intending to travel on the road, short of its actual termination. It will be readily seen that in some cases the traveller will be obliged to pay more than his fair proportion of toll for the distance travelled. While in other cases a traveller may pass over nineteen miles of the road by paying the toll for ten miles, and may also travel any distance between two gates, without paying any toll at all. It was the custom of the old turnpike companies by equitable arrangements for commutation and by prescribing reasonable low rates to be charged to those who were known to reside within a moderate distance from their gates, to avoid, as far as was practicable, the injustice of exacting full toll of those who had travelled but a short distance on the road. It is to be hoped that the plank road companies will adopt a regulation recommended alike by reason and justice. As to the particular question involved in this appeal we might stop here, and repose ourselves on the authority of the cases we have cited; for the language of the enactments to which a construction was given in those cases was far more favorable to the interpretation insisted on by the appellant's counsel than is the language of the act under consideration. But the legislature has given a construction to the provision in question entirely incompatable with that claimed by the counsel. By the second section of the act 'in relation to plank roads and turnpike roads' (Laws of 1849, P. 314, § 2, sub 5,) it is provided that 'persons living within one mile of any gate shall be permitted to pass the same at one half the usual rates of toll, subject to some exceptions mentioned in this act. Now it seems very clear that this must mean one half the amount of toll fixed as the toll to be received at the particular gate. It could not mean one half of the pro rata toll at a cent and a half per mile. That would be three fourths of a cent for a mile, and still more minute fractions of a cent for a less distance. The counsel for the appellant put some cases of flagrant injustice. That might occur under the law, upon the construction which we feel bound to adopt; as for instance, the location of but a single gate on the road, and that near the city of Utica, and an exaction of the toll for the entire route. We do not think such a case likely to occur; but if it should, we know of no remedy for that, and other like cases of importance, but an application for the removal of the gate; or such an amendment of the act regulating plank roads, as may reach the evil complained of. *Mallory vs Austin*, 7 *Bar. S. C. Rep.* 626.

§ 36. Upon filing such certificate as aforesaid, the company owning any turnpike road so inspected, may erect one or more toll gates upon its road, but not within three miles of each other, and may demand and receive toll not exceeding the following rates: For every vehicle drawn by one animal, three-quarters of a cent a mile; for every vehicle drawn by more than two animals, one and one-quarter cents a mile, and one-quarter cent additional a mile for every animal more than two; for every score of neat cattle one cent a mile; for every score of sheep or swine, one-half cent a mile; and in the same proportion for any greater or less number of neat cattle, sheep or swine; for every horse and rider, or led horse, one-half cent a mile: And in no case shall any such turnpike company charge or receive rates of toll which will enable it to divide more than twelve per cent. on its capital stock actually paid in, in cash, and invested in its road, after paying the expenses of managing the same, and keeping it in repair.

§ 37. The commissioners of highways of any town in which a toll gate may be located on any such road or in an adjoining town, whenever they, or a majority of them, shall be of opinion that the location of such gate is unjust to the public interest, by reason of the proximity of diverging roads, or for other reasons, may, on at least fifteen days written notice to the president or secretary of the said company, apply to the county court of the county in which such gate is located, for an order to alter or change the location of said gate. The court on such application and on hearing the respective parties, and on viewing the premises, if the said court shall deem such view necessary, shall make such order in the matter as the said court may deem just and proper; and either party may, within fifteen days thereafter, appeal from such order to the supreme court, on giving such security as said county judge shall require. Such order, unless appealed from, shall be observed by the respective parties, and may be enforced by attachment or otherwise, as the said court shall direct.—And if appealed from, the decision of the supreme court shall be final in the matter. The said county and supreme court may direct the payment of costs in the premises, as shall be deemed just and equitable.

§ 38. The business and property of such company shall be managed and conducted by a board of directors, consisting of not less than five nor more than nine, who, after the first year, shall be elected at such time and place as shall be directed by the by-laws of such corporation, and public notice shall be given of the time and place of holding such election, not less than twenty days previous thereto, in a newspaper printed in each county in or through which the road of such company is located. The election shall be made by such of the stockholders as shall attend for that purpose, either in person or by proxy. All elections shall be by ballot; and each stockholder shall be entitled to as many votes as he shall own shares of stock; and the persons having the greatest number of votes shall be directors. Whenever any vacancy shall happen in the board of directors, such vacancy shall be filled for the remainder of the year by the remaining directors. The directors shall hold their office for one year, and until others are elected in their places. No person shall be a director unless he is a stockholder in the company; and no stockholder shall be permitted to vote at any election for directors on any stock except such as he has owned for the thirty days next previous to the election.

§ 39. The directors of any company incorporated under this act may require payment of the sums subscribed to the capital stock, at such times, and in such proportions, and on such conditions as they shall see fit, ander the penalty of the forfeiture of their stock and all previous payments thereon; and they shall give notice of the payments thus required, and of the place and time when and where the same are to be made, at least thirty days previous to the payment of the same, in one newspaper printed in each county in or through which their road is located, or by sending such notice to such stockholder by mail, directed to him at his usual place of residence.

§ 40. The shares of any company formed under this act, shall be deemed personal property, and may be transferred as shall be prescribed by the by-laws of such company. The directors of every such company may, at any time, with the consent of a majority in amount, of the stockholders in such company, provide for such increase of the capital stock of such company as may be necessary to finish the making of a road actually commenced and partly constructed but the whole capital stock of any company shall not exceed five thousand dollars per mile for each mile of the road.

§ 41. It shall be the duty of the directors of every company formed under this act, to report annually, to the secretary of state, under the oath of any two of such directors, the cost of their road, the amount of all money expended, the amount of their capital stock, and how much paid in, and how much actually expended on such road; the amount received during the year, for tolls, and from other sources, stating each separately, the amount of dividends made, and the amount set apart for a reparation fund, and the amount of indebtedness of such company, specifying the object for which the indebtedness accrued.

§ 42. Within two weeks after the formation of any company, by virtue of this act, the directors thereof shall designate some place within a county in which, according to the articles of association of such company, its road, or some part thereof, is to be constructed, as the office of such company; and shall give public notice thereof, by publishing the same in a public newspaper, published in such county, which publication shall be continued once in each week, for three successive weeks, and shall file a copy of such notice in the office of the county clerk of every county in which any part of such road is constructed or is to be constructed. And if the place of such office shall be changed, like notice of such change shall be published and filed as aforesaid, before it shall take place, in which notice the time of making the change shall be specified. And every notice, summons declaration, or other paper required by law to be served on such company, may be served by leaving the same at such office with any person having charge thereof, at any time between nine o'clock in the forenoon and noon, and between two and five o'clock in the afternoon, of any day except Sunday.

§ 43. It shall be the duty of the directors of any such company to cause a book to be kept by the secretary, treasurer or clerk thereof, containing the names of all persons alphabetically arranged, who are or shall, within six years, have been, stockholders of such company, and showing their places of residence, the number of shares of stock held by them respectively, and the time when they respectively became the holders of such shares; which book shall, from nine o'clock in the forenoon until noon, and from two o'clock in the afternoon until five, on every day except Sunday and the fourth day of July, be open for the inspection of all persons who may desire to examine the same, at the office of such company, and any and every person shall have a right to make extracts from such book; and no transfer of stock shall be valid for any purpose whatever, except to render the person to whom it shall be transferred liable for the debts of the company, according to the provisions of this act, until it shall have been entered therein, as required by this section, by an entry showing to and from whom transferred. Such book shall be presumptive evidence of the facts therein stated, in favor of the plaintiff in any suit or proceeding against such company, or against any one or more stockholders, or against such company and one or more stockholders jointly. Every officer or agent of any such company, who shall neglect to make any proper entry in such book, or shall refuse or neglect to exhibit the same, or allow the same to be inspected and extracts taken therefrom, as provided by this section, shall be deemed guilty of a misdemeaner, and the company shall forfeit and pay to the party injured, a penalty of fifty dollars for every such neglect or refusal, and all the damages resulting therefrom. And every company that shall neglect to keep such a book open for inspection as aforesaid, shall forfeit to the people the sum of fifty dollars for every day it shall so neglect, to be sued for and recovered in the name of the people, by the district attorney of any county in or through which the road of such company shall be constructed, or shall be, according to its articles of association, intended to be constructed, and when so recovered, the amount shall be paid in equal portions to every county for the use thereof.

§ 44. The stockholders of every company incorporated under this act, shall be liable in their individual capacity for the payment of the debts of such company, for an amount equal to the amount of the stock they severally have subscribed or held in said company over and above such stock, to be recovered of the stockholder who is such, when the debt is contracted, or of any subsequent stockholder, and any stockholder who may have paid any demand against such company, either voluntarily or by compulsion, shall have a right to resort to the rest of the stockholders who were liable to contribution; and the dissolution of any company shall not release or effect the liability of any stockholder which may have been incurred before such dissolution.

§ 45. The debts and liabilities of any company under this act shall not exceed in amount, at any one time, fifty per cent. of the amount of its capital, actually paid in, and if such debts and liabilities shall at any time exceed such amount, the stockholders who were such at the time any excess of debts or liabilities shall be created or incurred, shall be jointly and severally individually liable for such excess, in addition to their other individual liability, as provided in this act.

§ 46. In any action against any company formed under the provisions of this act, the plaintiff may include as defendants any one or more of the stockholders of such company, who shall by virtue of the provisions of this act, be claimed to be liable to contribute to the payment of the plaintiff's claim; and if judgment be given against such company, in favor of the plaintiff for his claim or any part thereof, and any one or more of the stockholders so made defendants shall be found to be liable as aforesaid, judgment shall also be given against him or them, and shall show the extent of his or their liabilities individually. The execution upon such judgment shall direct the collection of the sum for which it may be issued, of the property of such company liable to be levied upon by virtue thereof; and in case such property sufficient to satisfy the same cannot be found in the county of the officer to whom the same shall be directed, that the deficiency, or so much thereof as the stockholders who shall be defendents in such judgment shall be liable to pay, shall be collected of the property of such stockholders respectively. And if in any such action, any one or more of such stockholders so found not to be liable, but no verdict or judgment in favor of any such stockholder shall prevent the plaintiff in such action from proceeding therein against the company alone, or against it and such defendants who are stockholders as shall be liable for such demand or some portion thereof. Suits may be brought against one or more stockholders who are claimed to be liable for any debt owing by the company, or any part of such debt, without joining the company in such suit; but no such suit shall be so brought until judgment on the demand shall have been obtained against the company and execution thereon returned unsatisfied in whole or in part, or the company shall have been dissolved; but it shall not be necessary that such dissolution shall have been declared by any judicial decree, sentence, or determination; and in such suit there may be a verdict and judgment in favor of any defendent not liable as aforesaid; but such verdict and judgment shall not prevent the plaintiff in such suit from proceeding therein against any defendant who shall be liable as aforesaid.

§ 47. Sections seven, eight nine, eleven, twelve, thirteen, fourteen, sixteen, twenty-two, twenty-three, twenty-four, thirty, thirty-five, thirty-six, forty-one, forty-two, forty-four, forty-five, forty-six, forty-seven, forty-eight, forty-nine, fifty-one and fifty-two of the first title of the eighteenth chapter of the third part of the Revised Statutes, shall apply to the companies organized by virtue of this act, and all inspectors and other officers named therein, and to all the officers and roads of such companies, so far as the same can be so applied and are consistent with this act.

§ 48. So much of any such road and of the toll

houses, gates and other appurtenances thereof constructed by virtue of this act, as shall be within any town, city or village, shall be liable to taxation in such town, city or village, as real estate.

§ 49. Every company incorporated under this act shall cease to be a body corporate:

1. If within two years from the filing of their articles of association they shall not have commenced the construction of their road, and actually expended thereon at least ten per cent. of the capital stock of such company, and,

2. If within five years from such filing of the articles of association such road shall not be completed according to the provisions of this act.

§ 50. All companies formed under this act shall at all times be subject to visitation and examination by the legislature, or by a committee appointed by either house thereof, or by any agent or officer in pursuance of law; and the courts of this state shall have the same jurisdiction over such corporations and their officers over those created by special acts.

§ 51. The legislature may at any time alter, amend or repeal this act, or may annul or repeal any corporation formed or created under this act.

§ 52. This act shall take effect immediately.

AN ACT *to amend the act to provide for the incorporation of companies to construct plank roads, and of companies to construct turnpike roads.*

Passed May 12, 1847.

The People of the State of New York, represented in Senate and Assembly, do enact as follows:

§ 1. The forty-seventh section of the act entitled "An act to provide for the incorporation of companies to construct plank roads, and of companies to construct turnpike roads, passed May 7, 1847, is hereby amended so as to read as follows:

§ 47. Sections seven, eight, nine, eleven, twelve, thirteen, fourteen, sixteen, twenty-one, twenty-two, twenty-three, twenty-four, thirty, thirty-five, thirty-six, forty-one, forty-two, forty-four, forty-five, forty-six, forty-seven, forty-eight, forty-nine, fifty, fifty-one and fifty-two, of the first title of the eighteenth chapter of the first part of the Revised Statutes, shall apply to the companies organized by virtue of this act, and all inspectors and other officers named therein, and to all the officers and roads of such companies, so far as the same can be so applied, and are consistent with this act.

§ 2. This act shall take effect immediately.

AN ACT *in relation to plank road and turnpike road companies.*

Passed November 24, 1847.

SEC. 1. Any company may procure by purchase or gift, lands for the construction of the road.
2. Survey of road to be made and recorded; the company to procure authority from supervisors of other counties through which the road may pass.
3. Where a company shall procure the land by this act, it shall have the same rights, &c., as under the general law.
4. Nothing herein shall authorize the bridging or obstructing streams, &c.

The People of the State of New York represented in Senate and Assembly, do enact as follows:

§ 1. Any company formed under the provisions of chapter 210 of the laws of 1847, entitled "An act to provide for the incorporation of companies to construct plank roads and companies to construct turnpike roads," may procure by purchase or gift from the owners thereof, any lands necessary for the construction of so much of its contemplated road as shall be intended to be constructed in any county. They may also procure by agreement from the officers named in the 26th section of said chapter, the right to take and use any part of any public highway necessary for the construction of so much of such road as shall be intended to be constructed in such county; and when any such company shall have procured all the lands necessary to be used for the construction of its road in such county, and the title to take and use such parts of the public highways in such county as shall be necessary for that purpose, such company may construct so much of its road as shall be intended to be constructed in such county, without making the application mentioned in the fourth section of the said chapter.

§ 2. Before proceeding to construct such part of its road as provided in the first section of this act, such company shall cause an accurate survey of such part to be made by a practical surveyor, signed by its president and secretary, acknowledged by them as conveyance of real estate, or to be acknowledged in order to be recorded in the county, and recorded in the same. It shall also, before proceeding to construct such part of its road, procure in manner provided by the said chapter, from the supervisors of every other county, if any there be, in which any portion of it is intended to be constructed, authority to construct the same through such other county; but in such case the commissioners appointed to survey and lay out the road of such company shall not be required to survey and lay out that portion of it intended to be constructed within the county, and which such company shall have procured the lands, and the right to take and use the public highways necessary for its construction as aforesaid.

§ 3. When any such company, by virtue of the provisions of this act, shall have procured the lands and the right to take and use the parts of any highways necessary to construct its road in any county, and shall have constructed the same without making the application mentioned in the fourth section of the said chapter, it shall possess the rights, powers and privileges and be subject to the same duties and liabilities in respect to its road and the part thereof so constructed, as if such application had been made and all the proceedings of such company had been had, pursuant to the provisions of the said chapter.

§ 4. Nothing in this act contained shall be deemed or construed to authorize the laying out or constructing of any road in the cases specified in section nine of chapter 210 of the laws of 1847, nor to authorize the bridging or obstructing of any stream navigable by vessels or steamboats.

§ 5. This act shall take effect immediately.

LAWS OF 1848.

AN ACT *to amend an act entitled "An act to provide for the incorporation of companies to construct plank roads, and companies to construct turnpike roads," passed May* 7, 1847.

Passed April 12, 1848.

The People of the State of New York represented in Senate and Assembly, do enact as follows:

§ 1. The commissioners appointed by the board of supervisors, as provided in the eighth section of the act to provide for the incorporation of companies to construct plank roads, and of companies to construct turnpike roads, passed May 7th, 1847, are hereby authorized in laying out a plank road to determine the distance that the outer limits of the road shall be apart, as they may judge necessary, provided, in no case shall the company take more than four rods in width, except by the voluntary sale of the same to the company.

§ 2. Any company formed under this said act, may take half the rates of tolls, and no more, provided for in said act, from persons living within one mile of the gate at which it is taken; but no tolls shall be taken from farmers going to and from their work on their farms.

LAWS OF 1849.

AN ACT *in relation to plank roads and turnpike roads.*

Passed April 6, 1849.

The People of the State of New York represented in Senate and Assembly, do enact as follows:

Section 1. The directors of any plank road company formed under the act passed May 7th, 1849, entitled "An act to provide for the incorporation of companies to construct plank roads, and for companies to construct turnpike roads," may with the written consent of a majority of the inspectors, whose appointment is provided for in the 23d* section of said act, construct branches to their main line of road, or extend their main line, or change the route of their road, or any part thereof, which branches or extensions shall in all respects be governed by the same rules and affected by the same laws as the main line of road, and the said directors may increase the capital stock of the company to an amount not exceeding two thousand dollars a mile of such branches or extensions for their construction, and distribute the certificates therefor among the stockholders of the company, in proportion to the stock owned by them severally, if such stockholders shall demand and pay up the same: and in case the new stock, after the directors have given public notice in some newspaper printed in every county in which their road is situated, for six successive weeks is not demanded and paid by the stockholders, they may permit any person or persons to subscribe and pay on the new stock the same per centage that had been paid on the original stock of the company, and the same shall in all respects be held and considered as though it had formed a part of the original stock of the company. The right of way for any such branches or extensions shall be acquired by the company in the same manner as is now provided by law for plank road companies to acquire the right of way for their roads.

§ 2. The following persons and no others, shall be exempt from the payment of tolls at the gates of the several plank road companies, formed under the before mentioned act:

1. Persons going to or from any court to which they have been summoned as jurors, or to which they have been subpœnaed as witnesses.

* This reference to the 23d section should be the 33d section. The error was probably made in writing out the bill by the clerk. The seventh section of this act makes the reference to the 33d section, as it should be.

2. Persons going to or from any training at which they are by law required to attend.

3. Persons going to or from religious meetings.

4. All persons going to or from any funeral, and all funeral processions.

5. Persons living within one mile of any gate, shall be permitted to pass the same at one-half the usual rates of toll, excepting farmers going to or returning from their work on their farms, who shall go free, when not employed in the transportion of persons or property of other persons.

6. Troops in the actual service of this state or of the United States.

7. Persons going to any town meeting or election at which they are entitled to vote, for the purpose of voting and returning therefrom.

§ 3. Any persons falsely representing him or herself to any toll gatherer, as being entitled to any of the exemptions mentioned in the preceding section of this act, shall forfeit to the company, to be recovered in the corporate name of the company, in any justice's court, the sum of ten dollars.

§ 4. Whenever any plank road company formed under the before mentioned act, shall have finished their road, or any three consecutive miles thereof, and had the same inspected as provided in the before mentioned act, it shall be lawful to erect a toll gate thereon, and exact toll thereat, for a term not exceeding one year, unless such road, or five consecutive miles thereof, be completed within such year.

§ 5. Sections fifty-four, fifty-five and fifty-six* of part first, title first, chapter eighteen of the Revised Statutes, shall apply to all companies formed under the before mentioned act, passed May 7, 1847, so far as the same can be applied, and are not inconsistent with this act.

§ 6. So much of section thirty-five of the aforesaid act, passed May 7, 1847, as provides that "in no case shall any plank road company charge or receive rates

* "Fifty-Six" repealed by act of March 16th, 1850.

of toll which will enable said company to divide, more nor shall any company divide more than ten per cent per annum on their capital stock actually paid in and invested in their road, after keeping the road in repair, and appropriating not exceeding ten per cent per annum on their capital stock invested as aforesaid, as a fund for the re-construction of their road when necessary," is hereby repealed.

§ 7. The inspectors, or a majority of them, whose appointment is provided for by the thirty-third section of the said act, passed May 7, 1847, are hereby authorized to determine the distance that the outer limits shall be apart, of any plank road or any turnpike road, belonging to any company formed under said act, in case the same has not been determined by the commissioners appointed under the eighth section of said act: provided, that in no case shall the company take more than four rods in width, except by the voluntary sale of the same to the company.

§ 8. It shall be lawful for any two or more companies, formed under the provisions of the aforesaid act of 1847, to consolidate the respective companies on such terms as the persons owning two-thirds of the stock of each of said companies may agree upon; and such company, consolidated as aforesaid, may change the name of their road, on filing in the office of the secretary of state, a certificate, containing the names of the roads so consolidated and the name by which such road shall thereafter be known.

§ 9. Any person who shall pass any plank road gate, or turnpike gate, without paying the toll, and with the intent to avoid the payment of the same, by which a penalty accrues, or any person committing any depredations or trespass on any plank road or turnpike, may be sued for said penalty or trespass in the county where such offence or trespass was committed, or in the county where such person may reside.

§ 10. Whenever a complaint shall be made to the inspector or inspectors, of any plank road or turnpike in this state, before such inspector or inspectors shall act upon such complaint, he or they shall receive from the complainant the fees provided by law, and in case it shall appear, upon examination of the road, that the complaint was well founded, the amount of said fees shall be paid to the complainant by the company. In case it is determined that the complaint was not well founded, the complainant shall not be entitled to receive back the fees so paid by him.

§ 11. Whenever any plank road or turnpike road, shall be built in pursuance of the provisions of this act, or the act hereby amended, upon the site of an old highway, it shall be the duty of the commissioners of the highways of the town where such road shall be made, to designate some district or districts within their town, on which the highway labor of the inhabitants residing along the line of said plank or turnpike shall be performed.

§ 12. It shall be lawful for the inhabitants residing in any road district in this state, to grade, gravel or plank the road or roads in such district, by anticipating the highway labor of such road district for one or more years, and applying it to the immediate construction of such plank or gravel road, the said inhabitants shall be exempted from the labor so anticipated and applied, except so far as their labor may be necessary to keep their said road or roads in repair; such road to be in all cases free road.

§ 13. Section first of chapter 201, of the laws of 1847, is amended by striking out in the tenth and eleventh lines, the words "and five per cent. paid thereon as hereinafter required."

PLANK ROAD ACT.

AN ACT *to amend the act for the incorporation of companies to construct plank roads and turnpike roads.* Passed March 16, 1850.

The People of the State of New York represented in Senate and Assembly, do enact as follows:

§ 1. Every company heretofore formed or organized under the act entitled "An act for the incorporation of companies to construct plank roads and of companies to construct turnpike roads," passed May 7, 1847, and the several acts amending the same shall be deemed to be a valid corporation, although such company may not have complied with the requirements of such acts in the formation and organization of such company; and no act or omission on the part of any such company, or of its stockholders or officers, shall work a forfeiture of its corporate powers or franchises, unless the same was wilful or malicious, but this section shall not affect or impair any right of action which has heretofore accrued to any person or persons against such company, its officers or agents for entering upon or taking possession of any real estate, or any right of action now existing arising on contract or any right of action against any company or its officers for a misapplication of its funds; or any action pending to recover toll or to recover any penalty for passing a gate without paying toll.

§ 2. Section five of the act entlitled "An act in relation to plank roads and turnpike roads," passed April 6, 1849, is repealed, and the following substituted in lieu thereof:

"§ 5. Sections fifty-four and fifty-five of title first, chapter eighteen of first part of the revised statutes shall apply to all companies under the before mentioned act, passed May 7, 1847, so far as the same can be applied or are not inconsistent with this act."

§ 3. No action to recover any penalty against any company formed under an act entitled "An act for the incorporation of companies to construct plank roads and of companies to construct turnpike roads," and the acts amending the same, or against any turnpike corporation, shall be commenced or maintained against such company or any of its officers or agents, unless the same is commenced within thirty days after the penalty was incurred.

§ 4. The second section of "An act in relation to plank roads and turnpike roads," passed April 6, 1849, is amended by adding thereto the following, to be the eighth subdivision of said section:

"8. All persons going to or returning from any grist-mill or blacksmith's shop, where they ordinarily get their grinding or blacksmith's work done, shall be exempt from the payment of toll at one gate only, within five miles of such person's residence, when he is going to or returning from such mill or shop for the express purpose of getting grinding or blacksmith's work done; but this exemption shall apply only to such parts of a plank road as have been or shall be made on a public traveled highway, not theretofore a turnpike."

§ 5. No supervisor or commissioner of highways of any town shall make any agreement with any plank road company or turnpike road company, under the first section of "an act in relation to plank road and turnpike road companies," passed November 24, 1847, for the right to take and use any part of any public highway for a plank road or a turnpike road without they first obtain the consent in writing of at least two-thirds of all the owners of land along such highway, who actually reside on that part of the highway on which such plank or turnpike road is to be constructed.

§ 6. No plank or turnpike road company shall hereafter erect or put up any hoistgate on their roads.

§ 7. This act, shall take effect immediately.

PLANK ROAD LAW.

AN ACT *in relation to plank roads and turnpike roads.*

Passed April 9, 1851.

The People of the State of New York represented in Senate and Assembly, do enact as follows:

§ 1. The following persons, and no others, shall be exempt from the payment of tolls at the gates of the several plank road companies, formed under the act entitled "An act to provide for the incorporation of companies to construct plank roads," passed May 7th, 1847:

1. Persons going to or from religious meetings, held at the place where such persons usually attend for religious worship, in the town where they reside, or an adjoining town, or within eight miles of their residence.

2. Persons going to or from any funeral, and all funeral processions.

3. Troops in the actual service of this State or of the United States, and persons going to or from a militia training, which by law they are required to attend.

4. Persons going to any town meeting or general election at which they are entitled to vote, for the purpose of voting or returning therefrom.

5. Persons living within one mile of any gate by the most usually traveled road, shall be permitted to pass the same at one-half the usual rates of toll, when not engaged in the transportation of other persons, or the property of other persons.

6. Farmers living on their farms within one mile of any gate by the most usually traveled road, shall be permitted to pass the same free of toll, when going to or from their work on said farms.

§ 2. It shall not at any time hereafter be lawful for any plank road company formed under the act of May seventh, eighteen hundred and forty-seven, or for any turnpike company to erect or put up any toll gate, gate house or other building within a less distance than ten rods from the front of any dwelling house, barn or other outhouse, without the written consent of the owner thereof; and if any toll gate or other such building shall hereafter be located by any such company within said distance, without such consent, the county judge of the county in which such building shall be so located, shall on application, order the same to be removed.

§ 3. Any thing contained in the act of May seventh, eighteen hundred and forty-seven, or in any subsequent act which is inconsistent with the provisions of this act, is hereby repealed.

REVISED STATUTES.

PART FIRST.—CHAPTER XVIII.—TITLE III.

Of the general Powers, Privileges and Liabilities of Corporations.

§ 1. Every corporation, as such, has power,

1. To have succession by its corporate name, for the period limited in its charter; and when no period is limited, perpetually:

2. To sue and be sued, complain and defend, in any court of law or equity:

3. To make and use a common seal, and alter the same at pleasure;

4. To hold, purchase, and convey such real and personal estate, as the purposes of the corporation shall require, not exceeding the amount limited in its charter:

5. To appoint such subordinate officers and agents, as the business of the corporation shall require, and to allow them a suitable compensation:

6. To make by-laws, not inconsistent with any existing law, for the management of its property, the regulation of its affairs, and for the transfer of its stock.

§ 2. The powers enumerated in the preceding section, shall vest in every corporation that shall hereafter be created, although they may not be specified in its charter, or in the act under which it shall be incorporated.

§ 3. In addition to the powers enumerated in the first section of this Title, and to those expressly given in its charter, or in the act under which it is or shall be incorporated, no corporation shall possess or exercise any corporate powers, except such as shall be necessary to the exercise of the powers so enumerated and given.

§ 4. No corporation created, or to be created, and not expressly incorporated for banking purposes, shall, by any implication or construction, be deemed to possess the power of discounting bills, notes, or other evidences of debt, of receiving deposits, of buying gold and silver, bullion, or foreign coins, of buying and selling bills of exchange, or of issuing bills, notes, or other evidences of debt, upon loan, for circulation as money.

§ 5. Where the whole capital of a corporation shall not have been paid in, and the capital shall be insufficient to satisfy the claims of its creditors, each stockholder shall be bound to pay on each share held by him, the sum necessary to complete the amount of such share, as fixed by the charter of the company, or such proportion of that sum, as shall be required to satisfy the debts of the company.

§ 6. When the corporate powers of any corporation are directed by its charter to be exercised by a particular body, or number of persons, a majority of such body, or persons, if it be not otherwise provided in the charter, shall be a sufficient number to form a board for the transaction of business; and every decision of a majority of the persons duly assembled as a board, shall be valid as a corporate act.

§ 7. If any corporation hereafter created by the legislature, shall not organize and commence the transaction of business within one year from the date of its incorporation, its corporate powers shall cease.

§ 8. The charter of every corporation, that shall hereafter be granted by the legislature, shall be subject to alteration, suspension and repeal, in the discretion of the legislature.

§ 6. Upon the dissolution of any corporation created, or to be created, and unless other persons shall be appointed by the legislature, or by some court of competent authority, the directors or managers of the affairs of such corporation at the time of its dissolution, by whatever name they may be known in law, shall be the trustees of the creditors and stockholders of the corporation dissolved, and shall have full power to settle the affairs of the corporation, collect and pay the outstanding debts, and divide among the stockholders the moneys and other property that shall remain, after the payment of debts and other necessary expenses.

§ 10. The persons so constituted trustees, shall have authority to sue for and recover, the debts and property of the dissolved corporation, by the name of the trustees of such corporation, describing it by its appropriate name, and shall be jointly and severally responsible to the creditors and stockholders of such corporation, to the extent of its property and effects that shall come into their hands.

Special Provisions relating to Certain Corporations.

§ 1. The book or books of any incorporated company in this state, in which the transfer of stock in any such company shall be registered, and the books containing the names of the stockholders in any such company, shall, at all reasonable times during the usual hours of transacting business, be open to the examination of every stockholder of such company, for thirty days previous to any election of directors; and if any officer having charge of such books, shall, upon demand by any stockholder as aforesaid, refuse or neglect to exhibit such books, or submit them to examination as aforesaid, he shall, for every such offence, forfeit the sum of two hundred and fifty dollars, ths one moiety thereof to the use of the people of this state, and the other moiety to him who will sue for the same, to be recovered by action of debt, in any court of record, together with the costs of such suit.

§ 2. It shall not be lawful for the directors or managers of any incorporated company in this state to make dividends, excepting from the surplus profits arising from the business of such corporation; and it shall not be lawful for the directors of any such company to divide, withdraw, or in any way pay to the stockholders, or any of them, any part of the capital stock, without the consent of the legislature; and it shall not be lawful for the directors of such company to discount or receive any note, or other evidence of debt, in payment of any instalment actually called in and required to be paid, or any part thereof, due or to become due on any stock in the said company; nor shall it be lawful for such directors to receive or discount any note or other evidence of debt, with the intent of enabling any stockholder in such company to withdraw any part of the money paid in by him on his stock; and in case of any violation of the provisions of this section, the directors under whose administration the same may have happened, except those who may have caused their dissent therefrom to be entered at large on the minutes of the said directors at the time, or were not present when the same did happen, shall, in their individual and private capacities, jointly and severally be liable to the said corporation, and to the creditors thereof, in the event of its dissolution, to the full amount of the capital stock of the said company so divided, withdrawn, paid out, or reduced, and to the full amount of any notes or other evidences of debt so taken or discounted in payment of any stock, and to the full amount of any notes or other evidences of debt with the intent aforesaid, with legal interest on the said respective sums, from the time such liability accrued; and no statute of limitatitations shall be a bar to any suit at law, or in equity, against such directors, for any sums for which they are made by liable by this section: *Provided*, That this section shall not be construed to prevent a division and distribution of the capital stock of such company which shall remain after the payment of all its debts, upon the dissolution of such company, or the expiration of its charter.

§ 3. The total amount of the debts which any incorporated company shall at any time owe, whether for deposits, or by bond, bill, note, or other contract, over and above the actual deposits with the said company, shall not any time exceed three times the amount of the capital stock actually paid in; and in case of any excess, the directors under whose administration the same may have happened, except those who may have caused their dissent therefrom to be entered at large on the minutes of the said directors at the time, and except those who were not present when the same did happen, shall, in their individual and private capacities, jointly and severally, be liable for such excess to the said corporation, and, in the event of its dissolution, to any of the creditors thereof, to the full amount of such excess, with legal interest from the time such liability accrued; and no statute of limitations shall be a bar at any suit at law, or in equity, against such directors, for any sums of money for which they are made liable by this section.

§ 4. Whenever any incorporated company shall have refused the payment of any of its notes, or other evidences of debt, in specie, or lawful money of the United States, it shall not be lawful for such company or any of its officers, to assign or transfer any of the property or choses in such company, to any officer or stockholder of such company, directly or indirectly, for the payment of any debt; and and it shall not be lawful to make any transfer or assignment in contemplation of the insolvency of such company, to any person or persons whatever; and every such transfer and assignment to such officer, stockholder or other person, or in trust for them or their benefit, shall be utterly void; and whenever any incorporated company shall have remained insolvent for one whole year, or for one year shall have neglected or refused to redeem its notes or other evidences of debt, in specie, or other lawful money of the United States, or shall for one year have suspended the ordinary business of such incorporation, such company shall thereupon be deemed and adjudged to have surrendered the rights, privileges and franchises, granted by any act of incorporation, and shall be deemed to be dissolved.

§ 5. It shall be the duty of the supreme court, upon the application of any person or persons, or body corporate, that may be aggrieved by, or may complain of, any election, or any proceeding, act or matter, in or touching the same, (reasonable notice having been given to the adverse party, or to those who are to be affected thereby, of such intended application,) to proceed forthwith, and in a summary way, to hear the affidavits, proofs and allegations of the parties, or otherwise inquire into the matters or causes of complaint, and thereupon to establish the election so complained of, or to order a new election, or make such order and give such relief in the premises, as right and justice may appear to the said supreme court to require: *Provided*, That the supreme court may, if the case shall appear to require it, either order an issue or issues to be made up in such manner and form as the supreme court may direct, in order to try the respective rights of the parties who may claim the same, to the office or offices of franchise in question; or may give leave to exhibit, or direct the attorney-general to exhibit, one or more information or informations in the nature of a quo warranto in the premises.

§ 6. No by-law of the directors and manages of any incorported company, regulatinig the election of directors of officers of such company, shall be valid, unless the same shall have been published for at least two weeks in some newspaper in the county where such election shall be held, at least thirty days before such election; and in all cases where the right of voting upon any share or shares of the stock of any incorporated company of this state, shall be questioned, it shall be the duty of the inspectors of the elections, to require the transfer books of said company, as evidence of stock held in said company; and all such shares as may appear standing thereon in the name of any person or persons, shall be voted on by such person or persons, directly by themselves, or by proxy, subject to the provisions of the act of incorporation.

§ 7. The inspectors who may be appointed to conduct any election of directors, or any other officer of any incorporated company of this state, shall be required, before entering on the duties of their appointment, to take or subscribe the following oath or affirmation: "I, A. B., do solemnly swear, [or affirm, as the case may be,] that I will execute the duties of an inspector for the election now to beheld, with strict impartiality, and according to the best of my ability."

§ 8. If at any time hereafter, the election for directors of any bank or other incorporated company of this state, shall not be duly held on the day designated and appointed by the act incorporating such bank or other incorporated company, it shall be the duty of the president and directors of such bank or other incorporated company, to notify and cause an election for directors to be held within sixty days immediately thereafter; and in all cases, no share or shares shall be voted upon, except by such person or persons who may have appeared on the transfer books of said company to have had the right to vote thereon, on the day when, by the act of incorporation of such company, the election ought to have been held: which said right so to vote shall be exercised by the persons so appearing as aforesaid upon the transfer books of such company, on any day when such election may be held.

§ 9. It shall not be lawful in any company incorporated for banking purposes, its officers, agents or servants, or any of them, directly or indirectly to purchase, or to be interested in the purchase of any promissory note, or other evidence of debt, issued by any such company, at a less sum than appears by the face thereof to be due and payable; and any person offending against the provisions of this section, shall forfeit and pay three times the nominal amount of the note or other evidence of debt so purchased, to be recovered with costs of suit, by any person who will sue for the same, in any court of competent jurisdiction.

§ 10. It shall not be lawful for any person being president, director, cashier, clerk, agent, or any way interested or concerned in the management of the concerns of any such company, to discount, or directly or indirectly make any loan upon any note, bill, or other evidence of debt, which shall have been offered to such directors for discount; and every note, bill, or other evidence of debt, so discounted, or upon which any loan shall have been made by any of the persons aforesaid, knowing that such note had been so offered and refused, shall be utterly void; and the person offending herein, knowing that such note had been so offered and refused, by making any discount or loan, shall, for every such offence, forfeit and pay to any person who will sue for the same, twice the amount of any such discount or loan, to be recovered by action of debt, with costs of suit, in any court of competent jurisdiction.

§ 11. The provisions of this Title shall not apply to any incorporated library, or religious society; nor to any monied corporation which shall have been or shall be created, or whose charter shall be renewed or extended, after the first day of January, one thousand eight hundred and twenty-eight, and which shall be subject to the provisions of the second Title of this Chapter.

TURNPIKE COMPANIES.

§ 7. An election for directors shall thereafter be annually held, on the same day of the same month on which the first election was held; and at each election, including the first, the stockholders present, by a plurality of votes, shall elect, by ballot, three persons, to preside at the next succeeding election.

§ 8. If an annual election shall not be held on the day fixed by law, it shall be held in the same manner, and with the like effect, on some early day, to be appointed by the directors then in office, who shall give and publish the same notice thereof, as is required in respect to the first election; and who after the day on which such election ought to have been held, shall be incapacitated from doing any act as directors, except such as may be necessary to give effect to the election so to be appointed.

§ 9. The persons presiding at each election shall, immediately after receiving the ballots, openly estimate the votes, and thereupon make and subscribe a certificate of the result. Of the first election, they shall make a return to the directors chosen, at their first meeting thereafter.

§ 11. Five directors shall be a board for the transaction of business, and the acts of a majority of the board shall bind the corporation.

§ 12. The directors, at their first meeting after their election, shall elect, by ballot, one of their number as president.

§ 13. The board shall supply every vacancy that may occur in the office of a director, and the person chosen shall hold his office until the next annual election. They shall also supply from the directors, every vacancy that shall occur in the office of president; and one of the members present shall be chosen by plurality of votes, to preside at every meeting of the board, from which the president shall be absent.

§ 14. The president and directors shall have power, and it shall be their duty,

1. To meet from time to time, at such place as they may deem expedient:

2. To make such by-laws, rules and regulations, as in their judgment, the affairs of the corporatiou shall require:

3. To appoint such subordinate officers, artists and workmen, as they shall deem necessary to execute the business of the corporation:

4. To continue to receive subscriptions of shares, until their whole capital stock shall be subscribed, unless it shall have been ascertained that a less sum will be sufficient to fulfil the ends of their incorporation:

5. To demand at such time and in such proportion as they shall see fit, from the respective stockholders, the sums of money due on their respective shares under pain of the forfeiture of such shares, and of all previous payments thereon, to the corporation.

6. To declare, by a by-law, in what manner, and under what restrictions, the shares of their capital stock shall be transferable:

7. To construct. complete, and keep in constant repair, the road, with all the necessary buildings and appurtenances, for the making of which they shall be incorporated:

8. To keep a fair and jnst account of all tolls received, and of all monies disbursed, and deducting costs and charges, to make and declare a dividend of the clear profits and income of the road, among the stockholders, on the first Tuesday of May and the first Tuesday of November, in every year.

9. To publish a notice of each dividend, in one or more of the public newspapers printed nearest to the route of the road, and of the time and place of the payment thereof, and to pay the same accordingly.

10. To report to the comptroller, within six months after the road shall be completed, an account of the expenses of its construction, and to exhibit annually to the comptroller, an account of the sums arising from the tolls, of the disbursement and of the dividends actually made within the year.

§ 16. Every such corporation may be dissolved by the legislature when by the income arising from tolls, it shall have been compensated for all monies expended in purchasing, making, repairing and taking care of its road, and have received in addition thereto, an average annual interest at the rate of ten per cent.; and on such dissolution, all the rights and property of such corporation, shall vest in the people of this state.

§ 21. A mile stone or post shall be erected and maintained by the corporation on each mile of the road, on which shall be fairly and legibly marked or inscribed the distance of such stone or post from the place of the commencement of the road; and when such road shall commence at the end of any other road, having mile stones or posts, on which the distance from any city or town is marked, a continuation of that distance shall in like manner be inscribed.

§ 22. A guide post shall also be erected at the intersection of every public road, leading into or from the turnpike, on which shall be inscribed the name of the place to which such intersecting road leads, in the direction to which the name on the guide post shall point.

§ 23. No director of the corporation to which it shall belong shall be concerned directly or indirectly in any contract for the making or working of the road, or any part thereof, during the time he shall be a director.

§ 24. No contractor for the making of such road, or any part thereof, shall make a new contract for the performance of his work, or any part thereof, other than by hiring hands, teams, carriages or utensils to be superintended and paid by himself, unless such new contract and its terms be laid before the board of directors, and be approved by them.

§ 30. Whenever an appraisement shall be made of the lands of any old road, used as such by prescription, on which a turnpike shall be laid out, the appraisers shall set down the value of the soil and of the improvements, and the monies paid by any town for making such improvements, in separate sums; and the sum for which the soil is appraised shall be paid to the owners thereof, and the value of the improvements, and the sums paid therefor, by any town, shall be paid to the commissioners of highways of the town in which such old road shall be situate.

§ 35, Each toll gatherer may detain and prevent from passing through his gate, the persons riding, leading or driving animals or carriages subject to toll, until they shall have paid respectively the tolls authorized by law.

§ 36. No tolls shall be collected at any gate of any company incorporated under this Title in either of the following cases:

1. From any person passing to or from public worship, or a funeral; to or from a grist-mill for the grinding of grain for family use; or to or from the blacksmith's shop to which he usually resorts for work there to be done:

2. From any person going for a physician or midwife, or returning from such errand; going to or returning from court when legally summoned as a juror or witness; going to or returning from a militia training which, by law, he is required to attend; or going to a town meeting or election at which he is entitled to vote, for the purpose of giving such vote, and returning therefrom:

3. From any person residing within one mile of the gate at which toll is demanded, unless he shall be employed in the carriage or transportation of the property of other persons, not so residing.

4. From troops in the service of this state, or of the United States.

§ 41. It shall be the duty of each inspector to whom a complaint in writing shall be made, that a turnpike road, or a part of such road in his county is out of repair, without delay to view and examine the road complained of; and if he shall find such claim to be just, he shall give notice in writing of the defect, to the toll gatherer, or person attending the gate nearest to each place out of repair, and in such notice, may, in his discretion, order such gate to be thrown open; but no inspector or inspectors shall order such gate to be opened unless a notice in writing shall have been served on the gate keeper nearest to the place out of repair, particularly describing such place at least three days previous to making such order.

§ 42. Immediately after the service of such notice, each gate ordered to be thrown open, shall be opened; nor shall it be again shut nor any toll be collected thereat, until one of the inspectors for the county shall have granted a certificate, that the road is in sufficient repair, and that such gate ought to be closed.

§ 44. Every keeper of a gate ordered to be thrown open, who shall not immediately obey such order, or who shall not keep open such gate until a certificate permitting it to be closed shall be granted, or who, during the time such gate ought to be open, shall hinder or delay any person in passing, or take or demand any tolls from any person passing, shall, for each offence, forfeit the sum of ten dollars to the party aggrieved.

§ 45. It shall be the duty of each inspector, who, upon due examination, shall have discovered a turnpike road, within his county, to be out of repair, or that any gate thereon is placed in a situation contrary to law, to give notice in writing of such defect or default, to one or more of the directors of the company to which such road shall belong.

§ 46. In such notice, he shall require the defective road to be repaired, or the gate improperly placed to be removed, within a certain time to be fixed in the notice; and, in his discretion, may order that, in the meantime, the gates on such road, or such of them as he shall specify, be thrown open.

47. If the requisitions of such notice be not obeyed, it shall be the duty of such inspector, to make immediate complaint to the attorney general, or the district attorney for the county, whose duty it shall be to prosecute the delinquent company, in the name of the people of this state. Such corporation, if convicted of having suffered their road to be out of repair, or having placed one or more of the gates thereon in a situation contrary to law, shall be fined in a sum not exceeding two hundred dollars.

§ 48. To each inspector of turnpikes, who shall view a turnpike road upon complaint made to him, shall be allowed the sum of two dollars for each day spent by him in the performance of such duty. If he shall adjudge the road viewed to be out of repair, such fees shall be paid by the company to which the road shall belong; otherwise, they shall be paid by the party making the complaint.

§ 49. Such fees, when payable by the company, shall be paid by the toll gatherer nearest the road adjudged out of repair, on demand and out of the tolls received or to be received by him; and may be recovered, with costs, of such toll gatherer, if he shall neglect or refuse to make such payment.

§ 50. Every toll gatherer who, at any turnpike gate, shall unreasonably hinder or dely any traveller or passenger liable to the payment of toll, or shall demand and receive from any person more toll than by law he is authorized to collect, shall. for each offence, forfeit the sum of five dollars to the person aggrieved.

§ 51. Whenever a judgment is obtained against a toll gatherer for a penalty, or for damages, for acts done or omitted to be done by him in his capacity of toll gatherer, and goods and chattels of the defendant to satisfy such judgment cannot be found, it shall be satisfied by the corporation whose officer he shall be; and if, on demand payment be refused by the corporation, the amount thereof may be recovered, with costs, of such corporation.

§ 52. The president and directors of every turnpike corporation created or to be created, may from time to time commute with any person, whose place of abode shall adjoin or be near to the road, for the toll payable at the nearest gate on each side of such place of abode; but no such commutation shall be for longer time than one year, and it may be renewed at the end of each period for which it shall be made.

APPENDIX.

The reference to sections in the highway act, are to the *first* edition and third of the Revised Statutes. Many changes have been made since the first edition, and some sections entirely repealed.

(No 1.)

Article 1. § 1. Subdivision 3.

Order amending Description of Road.

Whereas a road, used as a highway in the town of ***Portland***, *in the county of Chautauque, leading from the Erie road between the dwelling houses of Elijah Fay and John R. Coney, to the Old Erie Road on the east line of Elisha's Fay's farm,* was laid out by the commisioners of highways of the said town, on the 10*th day of June*, 1815, but not sufficiently described:

Now therefore, the undersigned, commissioners of highways of the said town, [*if all the commisioners meet but do not all sign the order, say* all the comissioners being present, *or if all do not meet, say* all the commissioners having been notified to meet at this time and place for the purpose,] having met *at the dwelling house of John Adams*, in the said town, for the purpose of causing said road to be ascertained, described and entered of record in the town clerk's office, and having caused a survey of the said road to be made, do order that the said road be and the same is hereby ascertained and described according to the said survey, beginning at [*here insert the description according to the survey.*]

A. B.,
E. F., } *Commissioners.*
C. D.,

Portland, July 1, 1851.

☞ The forms are prepared for three commissioners. Where there is but one commissioner, the form can be varied to conform to the fact.

(No. 2.)

ARTICLE 1. § 1. SUBDIVISION 3.

Order ascertaining and describing road used 20 years, but not recorded.

Whereas a road in the town of *Portland in the county of Chautauque, leading from the town line road, near the dwelling-housee of Ahira Hall, by the north line of the farm Lloyd Burr, to the old Erie road, east of the dwelling-house of Horace Adams*, has been used as a highway for twenty years, but not recorded:

Now therefore, [the rest like form No. 1.]

(No. 3.)

ARTICLE 1. § 1. SUBDIVISION 5.

Order of dividing town into road districts.

The undersigned, commissioners of highways of the town of *Arkwright in the county of Chautauque [here insert the attendance of, or notice to, all the commissioners, as in form No.* 1,] do hereby order that the said town be, and the same is hereby divided into *thirteen* road districts, in the following manner, viz:

District No. 1, shall comprise all that part of the said town lying north of the south line of the road leading from the dwelling-house of   , to the east line of said town, and all the inhabitants residing in said district, and all those residing on the said road above mentioned, liable to work on highway, and   and   and all persons residing with them on their farms, and liable to work on highways, are assigned the said district No. 1.

District No. 2, shall comprise, &c. [till the thirteen districts are described and as the inhabitants properly assigned to the several districts.]

Dated February 15, 1848.

A. B.,
C. D., } *Commissioners.*
E. F.,

(No. 4.)

ARTICLE 1. § 3.

Commissioners' Annual Account.

The commissioners of highways of the town of Auburn, in the county of Cayuga, render to the board of town auditors of the said town, the following account, viz:

1. The highway labor assessed in said town for the year ending the 28th day of February, 1848, was sixteen hundred and thirty-nine days, and the highway labor performed in said town during the said year, was fourteen hundred and sixty-seven days.

2. The said commissioners have received during the said year, the following sums of money for fines and commutations, and under roads and bridges in the said town, together with the probable expense thereof, viz:

The middle bridge across Owasco creek needs a new butment and new plank, the probable expense of which will be $85.06, to wit: for the butmennt $40.00, and for the plank, 2,250 feet, at $20.00 per M., $45.00. [State the other items with equal particularity.]

A. B., } C. D., } *Commissioners.* E. F., }

Auburn, Oct. 20, 1848.

(No. 5.)

UNDER SEC. 8.

New assessment by an Overseer, or two-thirds.

Whereas the quantity of labor assessed on the inhabitants of road district No. in the town of Ridgeway, in the county of Orleans, for the year 1845, deemed insufficient by the subscriber, the overseer of highways of said district to keep the roads in said district in repair:

I have therefore made another assessment on the actual residents in said road district in the same proportion, as near as may be, and not exceeding one-third of the number of days assessed in the same year, by the commissioners of highways of the said town of Ridgeway, on the inhabitants of said district. Which assessment is as follows:

A. C.	One day,	T. S.	Four days,
G. H.	Ten days,	M. S.	Three days,
O. P.	Eight days,	L. M.	One day.
R. S.	Ten days,		

Witness my hand at Ridgeway, this day of 1845.

JOHN JONES, *Overseer of District No.* ——

(No. 6.)

UNDER SEC. 4.

Statement and Estimate for the Supervisor.

To the Supervisor of the town of Manlius, in the county of Onondaga.

The commissioners of highways of said town, report that the following improvements (as in the 4th and 5th clauses of the annual report.)

(No. 7.)

Article 1. Sec. 12.

Assessment by Overseer for Scraper or Plow.

Whereas the commissioners of highways of the town of Ellicott, in the county of Chautauque, on the 10th day of April, 1848, by writing under their hands, directed and empowered me, *Henry Baker*, overseer of highways of road district No. 5 in said town, to procure a good and sufficient iron [*or steel*] shod scraper and a plow, [*or either separately,*] for the use of my said road district, to be paid for by the moneys arising from commutations and fines within such district; and whereas such moneys are insufficient for the purpose, by the amount of $8.50:

Now, therefore, I, the said overseer, according to the form of the statute in such case made and provided, do hereby assess the deficiency of eight dollars and fifty cents aforesaid, upon the inhabitants of the said district, in the proportion they are respectively assessed on the assessment roll of said town; which said assessment is as follows, viz:

Names of Inhabitants.	Town Assessment.	Overseer's Assessment
Abner Hazeltine,	$11 90	$1 19
Horace Allen,	17 90	1 79

A. B., *Overseer of Dist. No.* 5.

Dated April 15, 1848.

(No. 8.)

Under Sec. 4.

Appointment of Overseer in case of Vacancy.

Town of Nunda, Allegany Co., *ss.*

Whereas, a vacancy has occurred in the office of overseer of highways for road district No. of said town, by reason of the removal of John Smith, elected to said office from said town, (or refused to serve, death, as the case may be:)

Now therefore, by virtue of the power vested in us by the statute, in such case made and provided, we the undersigned, commissioners of highways of said town, (or in case *there are two in town, say,* all the commissionrrs of said town,) having met and deliberated on the subject embraced in the warrant, (or we the undersigned, two of the commissioners of highways of the said town of Nunda, having met and deliberated on the subject embraced in this warrant; all the commissioners of said town having been duly notified to attend said meeting of the commissioners to deliberate thereon,) do hereby, in order to fill said vacancy, appoint Joshua Thompson, overseer of highways of and for said road district No.   in said town of Nunda.

(No. 9.)

Under Sec. 17.

Complaint of Commissioners against an Overseer.

To the commissioners of highways of the town of Lisle, in the county of Broome.

The complaint of A. B., a resident of the town of Lisle, aforesaid, respectfully sheweth, that William Smith, the overseer of highways for road district No.   in said town, (has neglected and refused to warn *Thomas Williams, and*   *&c.*, to work on the highways in said district, after having been required to do so by the commissioners of highways of said town, or one of them; or has neglected to collect the sum of one dollar imposed as a fine upon   for neglect to appear and work on the highways in said district.) And the said A, B. hereby requires the commissioners of highways aforesaid, to prosecute the said William Smith, for said offence. A. B.

Dated,   1845.

(No. 10.)

Article 1. Sec. 17.

Security for prosecuting Overseer.

Whereas *Joseph Wait* has made complaint to the commissioners of highways of the town of *Ellicott*, that Henry Baker, overseer of highways of road district No. 5 in said town, has refused and neglected [*here insert the matter complained of, as in the complaint*;] Now, therefore, we, *Joseph Wait and William H. Tew*, do hereby covenant, promise and agree, with the said commissioners of highways, that we will well and truly indemnify and save them harmless against the costs which may be incurred in prosecuting for the penalty annexed to such refusal or neglect. Sealed with our seals, and dated the 20th day of June, 1840.

A. B., [L. S.]
C. D. [L. L.]

(No. 11.)

UNDER § 2, OF THE ACT OF 1832.

Notice to be given by Commissioners.

NOTICE is hereby given to the electors of the town of Franklinville, in the county of Cattaragus, that the Commissioners of highways of said town are of opinion, that the sum of two hundred and fifty dollars, as now allowed by law, will be insufficient to pay the expenses actually necessary for the improvement of roads and bridges in said town, and that the additional sum of $ is necessary to make a bridge across the creek, near the house of in said town, (*or* to repair the bridge, &c. or to improve the road at, &c.) And that we the undersigned, commissioners of said town, shall, at the next annual town meeting of said town, in open town meeting apply for a vote authorizing the said sum of $ to be raised for the purpose aforesaid.

A. B.,
C. D., } *Commissioners.*
Franklinville, 1846. E. F.,

(No. 12.)

UNDER § 21.

Overseer's list of persons liable to do highway labor.

I JOHN DOE, overseer of highways for road district No. of the town of Brutus, in the county of Cayuga, do certify, that the following is a true and correct list of all the inhabitants who are liable to work on highways in said road district No. in said town.

Brutus, June 1, 1845.

John Smith,	Richard Roe,
George Johnson,	Thomas Jones,

JOHN DOE, Overseer.

(No. 13.)

ARTICLE 2. § 22.

Commissioners' List and Statement of Non-resident Lands.

A list and statement of the contents of all lots, pieces or parcels of land, within the town of in the county of owned by non-residents therein :

Owners' Names.	Description of land.	Value according to last assessment roll.	No. of days.	Road district.
Jeremiah Newcomb.	E. Pt. L. 14, T. 5, Range 12.	$1,250 00.	7	5

, April 1, 1851. A. B., C. D., E. F., } *Commissioners of Highways.*

(No. 14.)

ARTICLE 2. § 24.

Commissioners' Estimate and Assessment of Highway Labor.

The undersigned commissioners of highway of the town of in the county of [*here insert the attendance of, or notice to, all the Commissioners, as in No.* 1,] having met at in said town, to ascertain, estimate and assess the highway labor to be performed in the said town the ensuing year, have ascertained, estimated, and assessed the same as follows, that is to say:

1. The whole number of days' work assessed for the year is *twelve hundred*, being at least three times the number of taxable inhabitants of the said town.
2. Every male inhabitant above the age of twenty-one years, (excepting ministers of the gospel and priests of every denomination, paupers, idiots and lunatics) being *four hundred and fifty-three*, is assessed one day [*or two days, or one day and a half, &c.*]
3. The residue of such work, being *seven hundred and forty-seven* days, is assessed as follows, viz:

Names.	No. of days.	Names.	No. of days.
A. B.	6	C. D.	5½

The lands in said town owned by non-residents, are assessed as follows, viz:

Names.	Description.	Value.	No. of days.
E. F.	N. Pt. L. 4, T. 3, Range 7.	$1,500 00.	6

, April 1, 1851. G. H., I. K., L. M., } *Commissioners of Highways.*

☞ The commissioners are to affix to the name of each person named in the list furnished by the overseers the number of days' work assessed upon such persons respectively, and add to the several lists the description and assessment of each tract of non-resident land in the respective road districts.

(No. 15.)

ARTICLE 2. § 24. SUBDIVISION 5, AND § 34.

Certificate of Commsssioners to Overseer's Lists.

We certify that the number of days affixed to each name and to each tract of non-resident land in the annexed list, is the true assessment made by us upon such persons and tracts respectively for the ensuing year.

A. B., } *Commissioners of*
C. D., } *Highways.*
, April 1, 1851. E. F., }

(No. 16.)

ARTICLE 2. § 35.

Overseer's Assessment of Persons left out of list, and of New Inhabitants.

The following named persons having been left out of the list of persons assessed to work on the highways in road district No. in the town of , in the county of , [*or having become inhabitants of road district No. in the town of , in the county of , since the list of assessments of highway labor for said district was made.*] Now I, , overseer of highways of said district, according to the form of the statute in such case made and provided, do hereby assess and rate the said persons in proportion to their real and personal estate, to work on the highways, as others rated by the commissioners on such list, subject to an appeal to the commissioners, which said assessment is as follows, viz:

A. B., four days.
C. D., three days.
, June 10, 1851. E. F., *Overseer.*

(No. 17.)

UNDER § 36.

Appeal to Commissioners from assessment of Overseers.

To the commissioners of highways of the town of Chazy, in the county of Clinton.

The undersigned having been assessed by the overseer of road district No. in said town days' labor on the highway, on the ground that he is a new inhabitant of said district, (or that his name had been omitted by the commissioners in said town,) and conceiving himself aggrieved by said assessment, doth hereby appeal from said assessment so made by said overseer to the commissioner of highways of said town.

Dated, Chazy, June 1846 A. B.

(No. 18.)

UNDER § 36.

Appeal to Court by non-residents.

TOWN OF CHATHAM, COLUMBIA CO., *ss.*

A. B. a non-resident owner of lands in said town, considering (*or* C. D. agent of A. B. a non-resident owner of lands in said town who considers) himself aggrieved in the assessment for highway labor by the commissioners of highways of said town upon the following described lands, to wit: (here insert the description as in the list or statement made by the commissioners,) DOTH hereby appeal from the assessment of said commissioners to the honorable , and three of the judges of the court of common pleas of the said county of Columbia.

Dated, Chatham, the day of 1846. A. B. or
A. B. by C. D. agent.

(No. 19.)

UNDER SEC. 41.

Commissioners' Consent.

Whereas A. B. a resident of road district No. in the town of Homer, in the County of Courtland, is assessed days' labor in district No. in said town, for lands therein situate, at his request we hereby approve of his applying the work assessed in respect to such lands in district No. where they are situated.

Homer, 1851.

(No. 20.)

UNDER SEC. 42.

Notice to Agent of Non-resident.

To C. D. agent of A. B. a non-resident owner of lands in the town of Delhi, in the county of Delaware Take notice, A. B. a non-resident of the said town is assessed days' labor in road district No. in said town, and that said labor is required to be performed on the day of next, and the days following, on the road between the house of and in said district.

Yours, &c. JOHN DOE, Overseer.

(No. 21.)

Under Sec. 43.

Notice in case of non-residents.

Notice is hereby given, that the labor assessed on the several tracts of land in the town of Fishkill, in the county of Dutchess, hereafter mentioned, which have been assessed as owned by non-residents, is to be performed on the day of next, on the highway of said district, between the dwelling houses of and and the owners of said land, or their agents, are hereby required to cause the said labor to be performed accordingly.

Owner's name.	*Description of land or tract.*	*Assessment.*
John Doe.	N. pt. of Lot No 50, 100 acres.	Ten days.

Dated, Fishkill, 1851. A. B. Overseer of District No.

(No. 22.)

Article 3. § 50.

Overseer's complaint for Idleness, Neglect, &c.

Seneca County, *ss.*

A. B., overseer of highways of road district No. in the town of Ovid, in the said county, being sworn, says that on the 17th day of June, 1848, he gave C. D., who resides in said district, and is assessed to work on the highways therein, notice to appear on the 19th day of June aforesaid, with a (*state what kind of team or implements were required,*] on the road, [*state where,*] to do such work; and that the said C. D. did not appear nor furnish any one in his stead, [*or did not bring such team or implement as was required. stating it: or when he so appeared, was idle, or hindered others from working—or whatever the complaint is,*] and has not paid the commutation money for said work, nor rendered a satisfactory excuse for such neglect.

A. B.

Subscribed and sworn to before me,
June 21, 1851.
G. H., J. P.

(No. 31.)

Article 3. § 51.

Summons.

Seneca County, *ss.*

To any constable of the town of Ovid, in said county:

Whereas A. B., overseer of highways of road district No. in the said town, has made complaint on oath before me, a justice of the peace of the said town, that C. D., a resident of said road district, and assessed to work on the highways therein, after being duly notified to appear on the 19th day of June, instant, with [*state what team or implements were required,*] to do such work; and that the said C. D., [*stating the matter of the complaint*]: and has not paid the commutation money nor rendered a satisfactory excuse.

You are therefore hereby required to summon the said C. D. to appear forthwith before me at my office in the said town, to shew cause why he should not be fined according to law for such refusal, [*or neglect, or misconduct.*]

G. H., J. P.

Ovid, June 21, 1851.

(No. 25.)

UNDER SEC. 51,

Conviction endorsed on the complaint.

TOWN OF ATTICA, Wyoming County, *ss.*

The within named Richard Roe, having been duly summoned to appear before me, J. B. the Justice within named, to show cause why a fine should not be imposed upon him for the offence set forth in the within complaint, and no sufficient cause having been shown by said Richard Roe; I do impose a fine of *one dollar*, (or whatever sum it may be,) upon the said Richard Roe, for the said offence, together with dollars and cents, for the costs of proceedings.

Witness my hand, this day of , 1851.

J. B——, *J. P.*

(No 26.)

UNDER § 51.

Constable's return on summons.

The within summons personally served on Richard Roe, by reading the same to him, this day of , 1851.

JOHN DOE, *Constable.*

Or, I have only served the within summons on Richard Roe within named, by leaving a true copy thereof, this day, at his personal abode.

June, 1851, JOHN DOE, *Constable.*

[No. 27.]

ARTICLE 3. § 52.

Warrant to levy Fine.

SENECA COUNTY, ss.

To any constable of the town of Ovid, in the said county:

You are hereby commanded to levy of the goods and chattles of C. D. *four dollars and eighteen cents;* being *one dollar* for a fine imposed by me, for [*specify the neglect or misconduct*], as set forth in the complaint of A. B., overseer of highways of road district No. , in the said town; and also *three dollars and eighteen cents,* for the costs of the proceedings on said complaint: and bring the said sum of money before me without delay.

Ovid, June 23, 1851. G. H., *J. P.*

[No. 28.]

Oversseer's list of non-resident lands, on which labor has not been performed.

The following is a list of all the lands of non-residents and of persons unknown, in the town of Bern, Albany county, which were taxed on my lists, on which the labor assessed by the commissioners of highways for the year 1848, has not been paid, and the amount of labor unpaid.

Owner's Names.	Description of Lands.	Assessed value.	Labor unpaid.
John Reed,........	S. W. Pt. L. 7, Range 14, T. 9. 75 acres.	$225 00	1½ days.

JOSEPH MILLS,
Overseer of Highways of road district
Berne, September 30, 1851. *No. 9, in said town.*

(No. 29.)

Affidavit on such List.

ALBANY COUNTY, ss.

Joseph Mills, overseer of highways in road district No 9, in the town of Bern, in said county, being sworn says, that he has given the notice required in the [thirty-third or thirty-fourth, or both, as the fact may be] section of Title 1, chap. 16, of the First Part of the Revised Statutes, and that the labor for which such land is returned has not been performed. JOSEPH MILLS.

Subscribed and sworn to before me, *September* 30, 1851. }

JACOB STOW, *J. P.*

(No. 30.)

Overseers Annual Account.

To the commissioners of highways of the town of Salem, Washington county.

The undersigned, overseers of highways of road district No. 11, in the said town, pursuant to law, renders the following account:

1. The names of all persons assessed to work on the highways in the district of which he is overseer.

Names.	No. of days.	Names.	No. of days.
John Williams,.....	3	Job Frost,.......	5

2. The names of all those who have actually worked on the highways, with the number of days they have so worked.

Names.	No. of days.	Names.	No. of days.
Job Frost,.......	5	John Williams,....	1½

3. The names of all those who have been fined, and the sums in which they have been fined.

Names.	Sums.	Names.	Sums.
John Williams,.......	$1 87½	Peter Jackson,.......	$0 75

4. The names of all those who have commuted, and the manner in which the moneys arising from fines and commutations have been expended by him.

Names.	Days.	Amount.	Names.	Days.	Amount.
James Johnson,	3½	$2 19	Ralph Clark,	1	$0 62½

Whole amount received for fines and commutations, as above stated, $5.44; which has been expended in [turnpiking the road from to , or in building a bridge across Muskrat brook, &c., as the fact is]

5. A list of all lands which he has returned to the supervisor for non-payment of taxes, and the amount of tax on each tract of land so returned [here copy the list delivered to the supervisor].

JOHN CRARY,
Overseers of Highways,
Salem, February 27, 1851. *District No.* 11.

Affidavit.

WASHINGTON COUNTY, SS.

John Crary, overseer of highways, of road district No. 11, in the town of Salem, in said county, being sworn says, that the annexed account is true.
J. CRARY.

Subscribed and sworn to before me, *February* 27, 1851.

STEPHEN ROSS,
Com. of Highways.

(No. 32.)

UNDER § 65.

Application for the alteration of a road.

To the commissioners of highways of the town of Hector, in the county of Tompkins:

The undersigned, residents of said town, or owning lands in said town and liable to be assessed for highway labor therein, do apply to said commissioners to alter the highway leading from the house of   to the   turnpike in said town as follows: (Insert a particular description of the proposed alteration, by known boundaries or objects, or courses and distances.) The proposed alteration passes through lands which are not improved, enclosed or cultivated, (or passes through the lands of John Doe and Samuel Johnson, who give their consent to said alterations.) T. R.
C. D.
Hector,   , 1851. E. F.

(No. 33.)

UNDER § 65.

Application to lay a new road over unimproved lands, or where the owner consent.

To the commissioners of highways of the town of Oswego, in the county of Tioga:

The undersigned persons, liable to be assessed for highway labor in said town, do hereby apply to the commissioners of highways of said town, to lay out a new road, of the width of *four* rods, through lands not enclosed, improved or cultivated, except that a part of said proposed road passes over the lands of R. S., who has consented to the laying said road, *and executed a written consent hereunto annexed,* (or has signified his consent by signing this petition,) as follows:—Beginning at &c. (Make a description, as in last.) O. S.
P. Q.
Owego,   , 1851. R. T.

☞ An application to discontinue a road will be substantially like the form No. 32.

(No. 34.)

UNDER § 68.

Order of commissioners to alter a highway.

At a meeting of the commissioners of highways of the town of Malone, in the county of Franklin, at   in the said town, on the day of 18 , all the said commissioners have met and deliberated on the subject embraced in this order, (*or if but two of the commissioners met, say,* all the said commissioners having been duly notified to attend the said meeting, for the purpose of deliberating on the subject matter of this order:) It is ordered and determined by the said commissioners, that a highway be laid out in the said town, "*upon application of A. B.*" (if there was an application,) and passing through the improved lands of John Doe and Samuel Johnson, who have consented thereto, (if there is any improved land taken,) said road commences near the easterly end of the bridge over creek, and running thence northerly up the said creek to the highway near the dwelling house of . The courses and distances of said road, according to a survey thereof, which the said commissioners have caused to be made, are as follows: Beginning, &c. (Insert the survey.) It is further ordered, that the above described line be the center of said highway, and that the said highway be of the width of *four* rods, (not less than three.)

In witness whereof, the said commissioners have hereunto subscribed their names, the day of , 1851.

A. B.,
C. D., } *Commissioners.*
E. F.,

(No. 35.)

UNDER SEC. 69.

Release by Owner.

A highway having been laid out, on the day of the date hereof, by the commissioners of highways of the town of Rome, in the county of Oneida, on the application of A. B., through the improved land of me, John Doe, commencing at &c. (insert the description of the route in the order:)

Now, know all by these presents, that I, the said John Doe, for value received, do hereby release all claim to damages by reason of the laying out and opening the said highway. Witness my hand and seal, the day of 1846. JOHN DOE, [L. S.)

(No. 36.)

UNDER SEC. 71.

Notice to be given by applicants for highway.

Notice is hereby given, that the subscriber (or subscribers) ha made an application to the commissioners of highways of the town of Seward, in the county of Schoharie, to lay out a highway in said town, commencing at , &c. (here give the description as in the application, and which highway will pass through the improved lands (or enclosed and cultivated, as the case may be,) of E. F., and that twelve freeholders of the said town will meet at on the day of instant, (or next,) at o'clock in the noon, to examine the ground through which the said highway is proposed to be laid. C. W.

Dated the day of , 1845.

[No. 36.]

ARTICLE 4. SEC. 78.

Application to County Court to appoint Commissioners to assess Damages.

To the Hon. the County Court of county.

Whereas the commissioners of highways of the town of in said county, by an order dated , 18 , have laid out a highway in said town, beginning [*insert description as in tha order:*]

Now therefore, the undersigned, commissioners of highways of the said town, hereby apply for the appointment of commissioners to assess the damages occasioned by the laying out of said highway, pursuant to the statute in such case made and provided.

F. G.
H. B. } *Commissioners.*
Dated June, 1851. N. R.

Order of Court thereon.

OSWEGO COUNTY COURT:

At a term of the said court, held at the court-house in the *city of Oswego*, on the 14th day of June, 1848.

PRESENT,

The Hon. ORLA H. WHITNEY,
County Judge, and
JOHN M. CHASEY and
LUNY THAYER, Esqrs.,
Justices of the Sessions.

On reading and filing the application of E. G., H. B. and N. R., Esqrs., commissioners of highways of the town of , in said county, setting forth the laying out of a highway in said town, beginning [*insert the description,*] and praying the appointment of commissions to assess the damages occasioned thereby; it is ordered that F. G., H. J. and L. M., be and they are hereby appointed such commissioners.

The order to be entered in the minutes of the court, and when copies are needed to be made, certified by the clerk as in other cases.

ARTICLE 4. § 78.

Assessment by Commissioners.

Whereas the undersigned, , and , were appointed by an order of the county court of the county of , made on the day of , 1848, an the application of E. G., H. B. and N. R., commissioners of highways of the town of , in said county, commissioners to assess the damages occasioned by the laying out of a highway in the said town, beginning [*here insert description as in the order*,] which highway passes through the improved lands of , and , and was laid out by the commissioners of highways of the said said town, by an order, dated , 1851.

Now therefore, we the said commissioners, having taken the oath, of office prescribed by the constitution, and having all met and acted on the matter committed to us, at the in said town, this day of , 1818, pursuant to a notice from said commissioners of highways, of at least six days, according to law and having taken a view of the premises, and heard the parties and such witnesses as have been offered before us; do thereupon, determine and assess the damages required to be assessed on the said highway, as follows, viz: We assess the damages of , at dollars. We assess the damages of , at dollars.

A. B.
C. D. } *Commissioners.*
E. F.

, July 30, 1851.

[No. 37.]

UNDER SEC. 73.

Freeholders' Certificate.

COUNTY OF JEFFERSON, TOWN OF BROWNVILLE. *ss.*

We, the undersigned, being freeholders of the town of Brownville, in said county, and not interested in the lands through which the highway hereinafter described is proposed to be laid, nor of kin to the owner thereof, having met on the day and date hereof, at in the said town, and having been first duly sworn, and having personally examined the route of the said proposed highway, and heard the reasons offered for or against the laying out a highway, pursuant to the application of E. F., commencing at, &c. (*here insert the description as in the application and notice;*) and which highway will pass through the improved (or enclosed, or cultivated) lands of the said C. D.

In witness whereof, we have hereunto subscribed our names, the day of , 1851.

[No. 37.]

ARTICLE 4. § 79.

Notice when the Commissioners or parties are aggrieved or dissatisfied.

Notice is hereby given that I , conceiving myself aggrieved, [*or we, the commissioners ot highways, feeling dissatisfied*] by the assessment of damages made by , , and commissioners appointed by the county court of the county of , to assess damages occasioned by the laying out of a highway in the town of , in said county, beginning [*insert description*] which said assesssment was filed in the office of the town clerk of the said town, on the day of , 1848, do hereby ask for a jury to re-assess the said damages, and such jury will be drawn at the clerk's office, of the town of , in said county, adjoining the said town of , on the day of , 1848, [*not less than* 10 *nor more than* 20 *days from the filing of the assessment*] by the town clerk of said town of , [*where drawn.*]
, June , 1851.

To be signed by the person aggrieved, or the commissioners dissatisfied.

[This notice must be served on the town clerk and opposite partd within 20 days from the filing of the assessment, and must be servey on the clerk first, and the other party within three days after; and all in season to have the drawing of the jury within the 20 days. For the manner of service see § 78.

[No. 38.]

UNDER SEC. 75.

Notice to Occupant.

TO MR. RICHARD ROE:

Sir:—Please to take notice, that we, the undersigned, the commissioners of highways of the town of Danesburg in the county of Schenectady, shall meet at in the said town, on the day of , 1851, at o'clock in the noon, to decide on the application made by A. B., to us to lay out a highway, commencing at , &c., (here give the description as in the application,) and which will pass through your enclosed (or improved, or cultivated) lands; twelve freeholders having certified that it is proper and necessary to lay out said highway.

Dated Danesburg, this day of , 1851.

Yours, &c.

A. B.,
C. D., } *Commissioners.*
E. F.,

[No. 38.]

ARTICLE 4. § 80.

Town Clerk's certificate of drawing of Jury.

I, , Town Clerk of the town of , in the county of , do hereby certify that the following twelve names were this day drawn by me from a box containing the names of all such persons now residents of said town, whose names are on the last list filed in the town clerk's office, of said town, of those selected and returned as jurors, pursuant to article 2, title 4, chapter 7, part 3, of the Revised Statutes, who are not interested in the lands through which a road in the town of , was laid out by the commissioners of highways of said last mentioned town on the day of , 1851, nor of kin to either or any of the parties, and that the said twelve names were so drawn by me to make a jury to re-assess the damages occasioned by the laying out of the said highway.

A. B. E. F.,
C. D. G. H., &c.,
, July , 1851. inserting the 12 names.
R. P., *Town Clerk.*

[No. 39.]

UNDER SEC. 76.

Order for laying out a highway through improved lands without the consent of the owner.

At a meeting of the commissioners of highways of the town of Lowville, in the county of Lewis, at in the said town, on the day of , all the said commissioners having met and deliberated on the subject matter of this order, (or if but two of the commissioners met, say, all the commissioners having been duly notified to attend the said meeting, for the purpose of deliberating on the subject matter of this order,) upon the application of A. B., a resident in said town, and liable to be assessed to work on the highways therein, for the laying out of the highway hereafter described, and on the certificate of twelve reputable freeholders of said town, convened and duly sworn after due public notice, as required by the Statute, certifying that such highway is necessary and proper; and notice in writing, of at least three days, having been given in due form of law to S. M., and R. S., occupants of the land through which such highway is to run, that the undersigned commissioners would meet at this time and place, to decide on the application aforesaid; and we having heard all reasons offered for and against laying out such highway; it is ordered, determined and certified, that a public highway shall be, and the same is hereby, laid out pursuant to said application, whereof a survey has been made, and is as follows, to wit: Beginning, &c., (as in the survey,) and the line of said survey is to be the center of said highway, which is to be rods in width.

Witness our hands, this day of , 1851.

A. B., }
C. D., } *Commissioners.*
E. F., }

[No. 39.]

ARTICLE 4. § 81.

Summons for Jury.

COUNTY, ss.

To , one of the constables of the town of in said county:

You are hereby directed to summon [*name the twelve persons*] to appear at , in the said town, on the day of , 1851, to make a jury to re-assess the damage occasioned by the laying out of a highway, in the said town, by the highway commissioners thereof, on the day of , 1851. Hereof fail not.

J. B., *J. P.*

(No. 40.)

ARTICLE 4. § 77.

Agreement of Owner and Commissioners, as to Damages.

Whereas the commissioners of the town of , in the county of , have, by an order dated the day of , 1851, laid out a highway in said town, beginning [*describe it as in the order,*] which said highway passes through the improved lands of J. H.: Now, therefore, the damages of the said J. H., by reason of the laying out of said highway, are hereby ascertained by agreement of the said J. H., and the said commissioners of highways, at the sum of *one hundred dollars.*

, July 18, 1851. J. H.

B. L., }
X. Z., } *Commissioners of Highways.*
P. M., }

[No. 41.]

ARTICLE 4. SEC. 82.

Oath of Jury.

You and each of you do solemnly swear, in the presence of Almighty God, well and truly to determine and re-assess such damages as shall be submitted to your consideration.

(No. 42.)

ARTICLE 4. § 82.

Verdict of Re-assessment.

We, the subscribers, a jury duly drawn and sworn, to determine and re-assess the damages occasioned by the laying out of a highway in the town of , in the county of , by the highway commissioners thereof, on the day of , 1851, which said highawy runs from [*describe the highway as in the order, and state whose lands it passes through,*] having taken a view of the premises and heard the parties and such witnesses as have been offered by them and sworn before us, do hereby determine and re-assess the said damages as follows, viz: We determine and re-assess the damages of H. B., at dollars, [*specify each person's damages passed upon.*]

To be signed by the six jurors.

, July , 1851.

COUNTY, ss.

I, J. B., the justice of the peace by whom the within [*or above*] named jury were summoned, drawn and sworn, do certify that the within [*or above*] is the verdict of re-assessment rendered by the said jury, pursuant to the said proceedings, this day of July, 1851.

J. B., *J. P.*

(No. 45.)

Application for a private Road.

To the commissioner [or commissioners] of highways of the town of Mendon in the county of Monroe:

The subscriber, a resident of said town, and liable to be assessed for highway labor, hereby makes application to you to lay out a private road for my use, beginning, &c. [insert description] passing through the lands of John Noble of said town.

Dated the day of , 1847.

(No. 46.)

Notice to the owner or occupant.

To of the town of Mendon, in the county of Monroe:

The subscriber having applied to the commissioners of highways of said town to lay out a private road, for my use, through your lands, [or lands which you occupy] you are hereby notified, that twelve disinterested freeholders will meet on the day of instant, at o'clock, M., at in said town, to be sworn and to proceed to view the lands through which the road is applied for, and to determine whether it is necessary. Dated, A. B.

(No. 47.)

Certificate of Jury, upon application for a private Road.

We, the undersigned, being disinterested freeholders of the town of , in the county of Erie, having met on the day of in the year , at , in said town, and having been duly sworn, well and truly to examine and certify with regard to the necessity and propriety of the road described in the annexed application of A. B., and having viewed the lands through which it is proposed to be made, do certify, that in our opinion it is necessary to and proper to lay out a private road for the use of the said A. B., pursuant to his said application.

In witness whereof, we have hereunto subscribed our names, this day of , 1845. P. W.
S. T., &c.

(No. 48.)

Order for laying out a private Road.

At a meeting of the commissioners of highways of the town of Mendon in the county of Monroe, at , in the said town, on the day of , all the said commissioners having met and deliberated on the subject of this order, [or if but two of the commissioners met, say, all the said commissioners having been duly notified to attend the said meeting, for the purpose of deliberating on the subject of this order,] upon the application of A. B. for the laying out of the private road hereafter described, and on the certificate of twelve reputable freeholders of said town, convened and duly sworn, after due notice to the owner [or occupant] of the lands through which said road is to pass, as required by the statute, certifying that such road was necessary: It is therefore ordered and determined by the said commissioners, that a private road be laid out for the use of the said A. B., pursuant to his application, the courses and distances whereof, according to a survey thereof which the said commissioners have caused to be made, are as follows: (insert the survey.) And it is further ordered, that the line above described, shall be the centre of said road, and that said road shall be of the width of rods.

In witness whereof, we we have hereunto subscribed our hands, this day of 1845.

A. B., | C. D., | E. F., | *Commissioners.*

(No. 49.)

Application to discontinue an old Road.

To the commissioners of highways of the town of Volney, in the county of Oswego:

The undersigned, a resident of said town, and liable to be assessed for highway labor therein, do hereby apply to you, the said commissioners, to discontinue the old road in said town, commencing near my dwelling-house, [or in any other suitable manner, giving an accurate description of the part sought to be discontinued,] on the ground that the said road has become useless and unnecessary.

Dated Volney, 1845. E. P.

(No. 50.)

Oath to Freeholders.

You, and each of you, do solemnly swear, that you will well and truly examine and certify, in regard to the propriety of discontinuing the road, for which application has been made by A. B

(No. 51.)

UNDER SEC. 86.

Freeholder's Certificate to discontinue a Road.

OTSEGO COUNTY, TOWN OF HARTWICK, *ss.*

We, the undersigned, disinterested freeholders of the the town of Hartwick, in the county of Otsego, having met at , on this day of 1845, in pursuance of the summons of the commissioners of highways of the said town, in order to examine and certify in regard to the propriety of discontinuing the road described in the annexed application of E. F , do certify, that in our opinion said road is useless and unnecessary, and that the same ought to be discontinued.

Ie witness whereof, we have hereunto subscribed our names, this day of , 1851.

G. H.
C. E. &c.

(No. 52.)

UNDER SEC. 86.

Order for Discontinuing a Road.

At a meeting of the commissioners of highways of the town of Carmel, in the county of Putnam, at in said town, on the day of , 1851, all the commissioners having met and deliberated on the subject of this order (*or* all the commissioners having been duly notified to attend the said meeting, for the purpose of deliberating on the subject of this order,) upon the application of A. B , of said town for the discontinuance of the road hereinafter described, and on the certificate of twelve disinterested freeholders duly summoned and sworn, who have in due form certified that said road is useless and unnecessary; and the said commissioners have caused a survey of said road to be made as follows, viz: (here insert the survey or description.) It is ordered and determined by the said commissioners, that the said road be, and the same is hereby discontinued. Witness our hands, this dated , 1851.

A. B.
C. D. } *Commissioners.*
E. F.

Order refusing to discontinue a Road.

At a meeting of the commissioners of highways of the town of Pittstown, in the county of Rensselaer, [as in last form.] It is ordered and determined by the said commissioners, that the said application of the said A. B. to discontinue the said road be refused.

In witness, &c.

A. B.,
C. D.. } *Commissioners of Highways.*
E. F.,

(No 53.)

ARTICLE 4. SEC. 102.

Appeal to the County Judge.

To the Hon. SCOTT LORD,
County Judge of Livingston county:

I, Holloway Long, conceiving myself aggrieved by the determination of the commissioners of highways of the town of York, in the said county, made on the 15th day of June, 1848, in [laying out, altering, discontinuing, refusing to lay out, alter or discontinue,] a highway in the said town, from [describe the road as in the order of the commissioners,] upon the application of John Rogers, do hereby appeal from the determination of the said commissioners, and pray the appointment of referees according to the form of the statute in such case made and provided, to hear and determine my said appeal.

The ground upon which my appeal is made, is that [here set forth the cause of complaint] and my appeal is brought to reverse entirely the determination of the commissioners which [or if part only, then.] And my appeal is brought only to reverse that part [or those parts] of the determination of the commissioners which [here set forth the particular part or parts.] A. B.

York, July, 1851.

(No. 54.)

Appointment of Referees.

LIVINGSTON COUNTY, *ss.*

Whereas, Holloway Long, of the town of York, in the said county has appealed from the determination of the commissioners of highways of the said town, made on the 15th day of June, 1848, in [laying out, altering, discontinuing, refusing to lay out, alter, discontinue,] a highway in the said town, which highway is particularly described in the said appeal hereto annexed, and whereas, Richard Bailey and John Smith have also appealed from the same determination of the commissioners, which said appeals are hereto annexed] and sixty days having elapsed after such determination has been filed in the office of the town clerk of the said town:

Now, therefore, I, Scott Lord, county judge of the said county, to whom the said appeal was [or appeals were,] addressed, according to the form of the statute in such case made and provided, do hereby appoint James Johnson, of the town of Naples, Hiram Denison, of the town of Lima, and James S. Wadsworth, of the town of Geneseo, three disinterested freeholders who have not been named by the parties interested in the appeal, and who are residents of the county, but not of the town wherein the road is located, as referees, to hear and determine all the appeals that have been brought in relation to the said determination of the said commissioners.

A. B., *County Judge Liv. Co.*

Mount Morris, Aug. 15, 1851.

(No. 55.)

Subpœna for Witness to attend and testify upon an Appeal.

Town of Ogden, Monroe County, *ss.* To and

You, and each of you, are hereby commanded, in the name of the people of the state of New York, to appear before the undersigned, referees, appointed by the county court, of Monroe county, at on the day of , at o'clock, in the noon, to testify respectively in a matter of a certain appeal made from the decision of the commissioners of highways of the town of to the said county court, and then and there to be heard on the part, and in behalf of E. F., appellant, (or to the said commissioners, as the case may be.)

Given under our hands, this day of , 18 .

A. B.
J. K. } *Referees.*
S. M.

[No. 56.]

Order of Referees deciding an Appeal.

Whereas, Holloway Long, of the town of York, in the county of Livingston, on the 1st day of July, 1848, appealed to the Hon. Scott Lord, county judge of the said county, from the determination of the commissioners of highways of the said town, made on the 15th day of June, 1848, in [*laying out, altering, discontinuing, refusing to lay out, alter, discontinue,*] a highway in the said town, which highway is particularly described in the said appeal hereto annexed, [*and whereas, Richard Bailey and John Smith, also appealed from the same determination of the said commissioners, which said appeals are also hereto annexed,*] and whereas, after the expiration of sixty days after such determination had been filed in the office of the town clerk of the said town, the said county judge, according to the form of the statute in such case made and provided, appointed us, James Johnson of the town of Naples, Hiram Dennison, of the town of Lima, and James S. Wadsworth, of the town of Geneseo, three disinterested freeholders, who had not been named by the parties interested in the appeal, and who are residents of the county, but not of the town wherein the road is located, as refeerees, to hear and determine all the appeals that had been brought in relation to the said determination of the said commissioners, which said appointment is hereto annexed, and we having given notice pursuant to law, to the said commissioners of highways, [*and to John Rogers an applicant for such road*] specifying the 25th day of August, 1848, as the time and the dwelling-house of Jacob Howe, in the said town of York, as the place, at which we would convene to hear the appeal, which notice was duly served at least eight days before the said time of convening, to wit, on the 16th day of August, 1848, and we having convened at the time and place specified in said notice, and, before proceeding to hear the said appeal, [*or appeals*] having been duly sworn by an officer authorized to take affidavits to be read in courts of record, to wit, Robert Jones, Esquire, a justice of the peace of the said county, faithfully to hear and determine the matters referred to us, have heard the proofs and allegations of the parties, and do thereupon order, determine, and adjudge that the said determination of the said commissioners of highways be, and the same is hereby reversed, or affirmed.

A. B.,
C. D., } *Referees.*
E. F.,

York, Aug. 25th, 1851.

(No. 57.)

Order of Referees when a road is laid out by them.

The same form as No. 45 to the word "*reversed.*"

And we do further order, determine, and adjudge, that in pursuance of the said application of the said John Rogers, a highway be, and the same is hereby laid out, beginning [*here insert description of the road, and if the description is by giving the centre line of the road state it so, and give the width of the road.*]

A. B.,
C. D., } *Referees.*
E. F.,

York, Aug. 25, 1851.

(No. 59.)

Notice to remove fences.

To Mr.

Whereas, the undersigned commissioners of highways, of the town of , in the county of , have laid out a public highway, by an order, dated , 1851, and duly filed with the town clerk of the said town which said road passes through enclosed lands, owned [*or occupied*] by you; and whereas our determination in the matter of laying out such road has not been appealed from; Now, therefore, please to take notice, that you are required to remove your fences from within the bounds of said highway, within sixty days after service hereof.

Yours, &c., A. B., C. D., E. F., *Commissioners of Highways.*

, Sept. , 1851.

(No. 60.)

Commissioners' order to remove encroachment.

Whereas, a highway was laid out in the town of , in the county of , on the day of , 1815, by the commissioners of highways, of the said town [*or by , and , three of the judges, &c., or referees, &c., as the fact is,*] beginning, [*insert description as in the order including a statement of the width of the road*] which road is encroached upon by the fence of , to the extent of [*state how much*] on the *north* side [*describe where.*]

Now, therefore, we, the commissioners of highways of the said town [*state the attendance of, or notice to, all the commissioners unless they all sign the order*] do hereby order that such fence be removed, so that such highway may be of the breadth originally intended.

F. G. H. B. N. R. *Commissioner of Highways.*

July 16, 1851.

(No. 61.)

ARTICLE 5. § 111.

Notice to remove encroachment.

To Mr.

Sir—Please to take notice that an order, of which the annexed is a copy, has been duly made by the commissioners of highways of the town of , in the county of , and that you are hereby required to remove the fence therein specified, as encroaching upon the highway, within sixty days after service hereof.

N. R., P. Y., D. M., *Commissioners of Highways.*

, July 16, 1851.

(No. 62.)

Summons to freeholders in case of encroachment.

ONEIDA COUNTY, *ss.*

To any constable of the town of , in the said county.

You are hereby commanded to summon twelve freeholders of said town to meet on the day of , at the dwelling house of , in said town, to inquire into the matter of an alleged encroachment upon the highway in said town leading from [*here describe the place and the alleged encroachment, as in the order* and you are to give at least three days' notice to the commissioners of highways of said town and to , of the time and place at which such freeholders are to meet.

J. W.
Justice of the Peace.

June 1851

(No. 63.)

Oath to jury.

You and each of you do solemnly swear, that you will well and truly inquire whether any encroachment has been made, and by whom, on the highway now in question.

Oath to witness.

You do swear, that the evidence you shall give, in relation to the encroachment on the highway now in question, shall be the truth the whole truth and nothing but the truth.

(No. 64.)

Certificate of Jury.

ONEIDA COUNTY, *ss.*

The undersigned, freeholders, of the town of , in said county, having met on the day of the date hereof, at the dwelling house of , in said town, pursuant to a summons from , Esq., a justice of the peace of the said county, to inquire into the matter of an alleged encroachment on the highway, in said town, specified in an order of the commissioners of highways of said town, dated , 18 , a copy whereof is hereto annexed, having been duly sworn, and having heard the proofs and allegations which were submitted, do certify that an encroachment on the said highway has been made, and that the same was made by , the present occupant, [*or by Samuel Smith a former occupant.*]

And the particulars of such encroachment are as follows, viz: [*here describe the encroachment, with certainty as to its location and extent.*]

To be signed by the freeholders.

, June 20, 1851.

(No. 64.)

Warrant for collection of Costs.

Oneida County, ss.

To any constable of the town of Camden, in the said county:

You are hereby commanded to levy of the goods and chattles of *John Jones,* of the said town, the sum of *four dollars and eighty-eight cents,* for the costs of an inquiry into the matter of an encroachment on the highway in said town, [*specify the place*] which said encroachment was certified to have been made by the said John Jones, [or by Samuel Smith, a former occupant of the land where such encroachment is made, which is now occupied by said John Jones,] by twelve freeholders of the said town on the 20th day of June, 1848, the costs of which inquiry are settled and adjudged by me at the sum above mentioned, and bring the said sum of money before me without delay.

A. D.
Justice of the Peace.

Camden, July 1, 1851.

(No. 65.)

When no encroachment is found.

Follow the form in No. 64 to the words, Do certify; then add—

That no such encroachment has been made on the public highway, and we have ascertained and do certify the damages of J. R. the occupant of the land through (or by) whose lands said highway runs, by reason of the said commissioners' proceedings against him, to be dollars.

(No. 66.)

Statement of Overseer in relation to gates.

To the Town Clerk of Attica, in the county of Wyoming.

The following is a true and correct statement of the charges incurred in the erection (or in repairing) of a gate in highway district

No. in said town, where the said highway runs through lands of liable to be overflowed by the waters of brook (or creek.)

1847. April 1. Paid for 100 feet of boards, $1 00
Paid A. B. for———, 2 00

Given under my hand, at Attica, this day of , 1846.

E. F. Overseer of district No.

WYOMING COUNTY, Town of Attica, ss.

E. F. being duly sworn, deposes and says that the above statement and account is true and correct.

Sworn and subscribed, &c. E. F.

(No. 67.)

Notice of an application for a ferry.

To John Smith. Sir—You will please to take notice, that an application will be made by me to the Court of Common Pleas of the county of Allegany, at the next term of said Court, to be holden at the Court House in Angelica, on the first Monday of next, for a licence to be granted to me to keep a ferry across the Genesee River in the town of in said county, where (here describe the place where it is proposed to have the ferry.)

Yours &c.

RICHARD ROE.

Dated , 1847.

(No. 68.)

Bond.

Know all men by these presents, that I, William Fellows, John Doe and Richard Roe, of the town of Penfield in the county of Monroe and state of New York, am held and firmly bound unto Daniel E. Lewis, Supervisor of said town of Penfield in the penal sum of one thousand dollars, to be paid to the said Daniel E. Lewis, or to his successor in office. For the payment of which we bind ourselves, our heirs and legal representatives jointly, severally and firmly by these presents.

Sealed with our seals and dated: &c.

Whereas, the above bounden William Fellows has been chosen Commissioner (or one of the commissioners) of highways of said town, at the late annual town meeting of said town (or has been appointed by two of the justices of the peace of said town to fill a vacancy.)

Now the condition of this obligation is such, that if the above bounden William Fellows, shall faithfully discharge his duties as such commissioner and within ten days after the expiration of his term of office, pay over to his successor, what money may be remaining in his hands as such commissioner, and render to such successor a true account of all moneys received and paid out by him as such commissioner, then this obligation to be void, otherwise to remain in full force and virtue.

Signed, Sealed and delivered in presence of } [L. S.] [L. S.] [L. S.]

Supervisor's Approval.

I approve of the within Bond. Dated , 185 .

DANIEL E. LEWIS,
"Filed March , 185 ." *Supervisor.*

FORMS

UNDER THE PLANK ROAD AND TURNPIKE LAW.

[No. 69.]

SEC. 1. PAGE 111.

Notice to open Books, &c.

BRIGHTON PLANK ROAD.

NOTICE is hereby given that books for subscribing to the Capital Stock for a Plank Road [*or turnpike*] to be constructed under the act of May 7, 1847, to run from the city of Rochester to the village of Pittsford, in the county of Monroe, through the village of , will be opened at the Eagle Tavern in said city, on the day of , 185 , at 10 o'clock A. M., and continue open from that time till 4 P. M., on that and each succeeding day for days.

A. B.,
Dated , 1851. C. D., &c.

☞ This notice must be printed in one or more newspapers according to § 1., and may be signed by one or more persons.

(No. 70.)

SEC. 1. PAGE 111.

Books of Subscription.

BRIGHTON PLANK ROAD.

To form an association under the Plank Road Act, so called, passed May 7, 1847, in pursuance of public notice, with a capital of thousand dollars, divided into shares of $ each, the subscribers severally and mutually agree to take and pay for the shares of stock set opposite our names, to the directors of the company hereafter to be formed under said act, in such sums and at such times and places as such directors may order, for the purpose of constructing a Plank Road from the city of Rochester to the village of Pittsford, in the county of Monroe.

Dated, &c.

Names.	Residence.	No. of Shares.

(No. 71.)

SEC. 1.

Notice to Subscribers, of Election of Directors.

BRIGHTON PLANK ROAD.

NOTICE is hereby given, that the subscribers to the stock of this road will meet at , on the day of , 185 , at 12 M., to elect directors for said company.

Dated, &c., C. D.

SEC. 1.

Articles of Association for the Brighton Plank Road Company.

ARTICLE 1. This association shall be called and known as the "Brighton Plank Road Company."

2. The said company is to continue for the term of years from the date hereof.

3. The amount of capital stock of said company shall be thousand dollars.

4. The said capital stock shall consist of hundred shares of dollars each.

5. The concerns of said company shall be managed by directors, to wit: A. B.; C. D.; E. F.; G. H.; and J. K.; [not less than five nor more than nine,] until others are chosen according to law.
6. The said road shall be constructed from to [, and shall pass through (the vilage of [,) and shall be, as nearly as can be estimated, miles and rods in length.
7. The stockholders owning in good faith a majority of all the capital stock of this association, may, at any legally notified meeting of the members thereof, pass such by-laws for the regulation and management of the concerns of the same, as they shall deem necessary, in accordance with the law.

Dated, &c.

Names of Subscribers.	Place of Residence.	No. of Shares of Stock.
A. B.,	Pittsford.	Twenty-five shares.

[No. 72.]

Affidavit of Directors required by § 2.

MONROE COUNTY, *ss.*

A. B., C. D., E. F., Directors of the Brighton Plank Road Company, being severally duly sworn, say, that the amount of capital stock required by the first section of the "Plank Road Act," so called, passed may 7, 1847, to wit: the amount of at least five hundred dollars for every mile of the road intended to be built by said company—has been in good faith subscribed; and that five per cent. on the amount thus subcribed has actually been paid, in cash, to the directors named in the said articles of association.

Sworn, &c.

A. B.,
C. D., } *Directors.*
E. F.,

[No. 73.]

Application to Supervisors for authority to lay out the Road.—§ 4.

The Brighton Plank Road Company, hereby apply to the Board of Supervisors of the county of , for authority to lay out and construct a Plank Road, under the act of May 7, 1847, from to , passing through the village of , and to be, as near as can be estimated, miles in length.

Dated, &c. A. L., *Secretary.*

(No 74.)

Notice of Application to Supervisors.—§ 4.

Notice is hereby given, that the foregoing application [attach the notice to the application,] will be presented to the board of Supervisors of Monroe county, on the day of , 18 ; that the proposed plank road is to be constructed of a single track of eight feet, or thereabouts, in width, and to run from the city of , and pass through the village of , and the town of , to the village of A. B., *Sec. B. P. R. Co.*

Dated, &c.

(No. 75.)

Notice of three Supervisiors for special Meeting of the Board,—§ 5.

The undersigned, supervisors of Brighton, Rush and Henrietta, in the county of Monroe, hereby give notice, that a meeting of the board of supervisors of said county, will be held at the Mansion House in the city of Rochester, on the day of , at 10 o'clock, A. M., to hear an application of the Rush Plank Road Company, for authority to lay out and construct a Plank Road from to , passing through , and to take and hold the the real estate necessary for such purpose.

Dated, &c.

O. P.
T. U. } *Supervisors.*
R. S.

(No. 76.)

Affidavit of Service of the foregoing Notice.

ONTARIO COUNTY, *ss.*

A. K., of Phelps, being duly sworn, saith that he personally serqed the within notice on A. M., the Supervisor of , by reading the same (or stating the substance thereof, or that he served it by leaving a copy at his place of residence,) to him, on the day of , 18 ; and that service was in like manner made of the same notice on C. D., the supervisor of , on the day of , aforesaid. (And so for any other service.)

Sworn, &c. S. L.

This proof of service seems necessary to make the meeting legal.

13

(No. 77.)

Oath of Commissioners appointed under § 8.

"I do solemnly swear, (or affirm, as the case may be) that I will support the Constitution of the United States, and the Constitution of the State of New York; and that I will faithfully discharge the duties of a commissioner to lay out a plank road for the Brighton Plank Road Company, from the town to the , according to the best of my ability."

If the road is intended to run into or through more than one county commissioners must be appointed on application to the supervisors in each county, in the same manner. § 8.

(No. 78.)

Petition to County Judge.—§ 12.

To the Hon. P. G. Buchan, Monroe County Judge:

The petition of the Brighton Plank Road Company, by their secretary, humbly showeth—that the route surveyed and established by the commissioners appointed to lay out the road intended to be constructed by said company, runs in part through lands of M. B., in the town of ; and your petitioners further show, that the said company cannot agree with the said M. B., for the purchase of said lands, (or that he is a non-resident, insane or an idiot or an infant, and other reason, and what means have been taken to ascertain his name or residence,) which lands are described as follows, [describe the lands.] Your petitioners therefore pray, that the compensation and damages of the owner (or owners) of said lands may be ascertained by a jury. A. B., *Sec.* B. P. R. Co.

Dated, &c.

(No. 79.)

Affidavit of Verification.—§ 12.

Monroe County, ss.

A. B., C. D., and E. F., Directors of the Brighton Plank Road Company, being severally duly sworn, say, that the facts set forth in the annexed [or within] petition are true. A. B.

C. D.

Sworn, &c. E. F.

(No. 80.)

Affidavit that name or residence of owner is unknown.—§ 12.

MONROE COUNTY, ss.

I. M., of the city of Rochester, being duly sworn, saith, that the owner of the lands described in the annexed petition is unknown to this deponent, and his residence or name cannot, after careful inquiry, be ascertained by him. [State any facts to satisfy the judge.]

Sworn &c. I. M.

(No. 81.)

Order of Judge—§ 13.

Ordered on petition of the Brighton Plank Road Company,—That a jury be drawn according to law at the clerk's office of Monroe county, on the day of , 185 , at 1 o'clock, P. M., to ascertain the compensation and damages to A , B and C , owners of lands through which the said company intend to construct their plank road.

P. G. B.
Monroe Co. Judge.

(No. 82.)

Notice upon such Order.—§ 13.

Notice is hereby given, that, in pursuance of the annexed order, [copy the order to annex,] a jury will be drawn at the time and place therein set forth, and as required by the statute in such cases.

A. B., *Sec.* B. P. R. Co.

Affidavit of service of the Notice in § 13.

MONROE COUNTY, ss.

C. L., of Brighton, being duly sworn, saith, that he personally served [or otherwise, as the case may be,] the notice of which the annexed is a copy, on A , B and C , therein named, on the day of , 185 .

Sworn, &c. S. T.

(No. 84.)

Appointment of Guardian.—§ 14.

I hereby appoint J. K., of as a suitable person to take care of the interests of L. M., an infant (or idiot or married woman, as the case may be,) owner of certain lands intended to be taken and used by the Brighton Plank Road Company, in the construction of their plank road, in respect to the proceedings to ascertain his compensation and damages by a jury according to law.

P. G. B.
Monroe Co. Judge.

(No. 85.)

Affidavits of Service.—§ 17.

COLUMBIA COUNTY ss.

Sheriff, or (one of the constables) of said county being duly sworn says, that he personally served the annexed precept on A. and B. jurors in said precept named on the day of 18 , (or left at their residence, a notice containing the substance &c.,) and the distance he necessarily travelled for that purpose was miles. And in like manner (or how) and on the same day, on D. also a juror therein named, and the distance &c., (state how the service was made on all, and the distance.)

Sworn &c. A. B.

[No. 86.]

Subpœna for Witness.—§ 19.

MONROE COUNTY, ss. To Greeting:—

You are hereby commanded in the name of the people of the State of New York, to appear and attend at in the Town of on the day of 18 , before a jury, then and there to be empannelled, to testify before such jury what you may know in respect to the damages of A. (or B.) in consequence of a Plank Road intended to be constructed on or through the lands of said A. by the Brighton Plank Road Company, and then and there to be inquired into and determined by said jury.

Given under my hand at on the day of 18

A. B.. Justice of Peace.

[No. 87.]

Order, fixing the Time and Place of Meeting of Jury.—§ 20.

On petition of the Brighton Plank Road Company it is ordered, that the jury to ascertain the compensation and damages of A. B. C. and D. for their lands intended to be occupied in laying out and constructing the Plank Road of said company, meet for such purpose at the House of the village in on the day of 18 , at 11 o'clock A. M.

Dated &c. P. G. B.
Monroe Co. Judge.

[No. 88.]

Notice to Owners of meeting of Jury.—§ 20.

Take notice that the jury to ascertain the compensation and damages of certain persons in the annexed order mentioned (attach the notice to a copy of the order) for lands intended to be occupied by the Brighton Plank Road Co. in laying out and constructing their road, will meet at the time and place in the said order set forth.

Dated &c. A. B
To A. B. Sec'y. B. P. R. Co.

[No. 89.]

Affidavit of Service of the Foregoing Notice.—§ 24.

MONROE COUNTY, ss.

J K. of in said county being duly sworn, saith, that on the day of 18 he served a notice of which the within is a copy (attach a copy of the notice) on A. B. therein named by delivering the same to him personally (or leaving the same at his residence in) and further that he served the said notice on C. L. also in said notice named, by putting the same enclosed into the post office in , directed to him at his place of residence in and paying the postage thereon, and further that (state how the service was made upon any other owner in or out of the state, and whether it was published as required by Section 20.)

Sworn &c. J. K.

(No. 90.)

Verdict of the Jury.—§ 23—25.

In pursuance of an order of Wayne County Judge, bearing date the day of 18 , a copy of which is hereto annexed, the undersigned jurors drawn and summoned to ascertain and determine the compensation and damages that ought to be paid to C. L. D. T. any J. Y. owners of land to be taken by the Palmyra Plank Road Company, and for taking the same for said road, and also the amount that ought to be paid to them for the time spent and the necessary expenses incurred in respect to the proceedings, having met at the time and place in said order mentioned, and heard all the proofs and allegations of the parties, and viewed the lands in question, have ascertained and do determine upon agreement thereupon, and therefore certify, that the said Company ought to pay to the said C. S. for his land thus taken to wit:—[describe it particularly] together with his time spent and the necessary expenses incurred by him in respect to the proceedings, the sum of dollars and cents. To the said D. T. for his land thus taken to wit: [describe it] and his time and necessary expenses incurred as aforesaid, the sum of dollars and cents. [And in the same manner for all whose lands are taken. (At least 12 must sit.)

Dated &c.

A. B. }
E. B. } Jurors.
G. B. }

[No. 91.]

Agreement with the Supervisor and Commissioners.—§ 26.

It is hereby agreed by and between the Supervisor and Commissioners of highways of the Town of Sandlake in the county of Rennsselaer on the one part, and the Stephentown Plank Road Company on the other part, that the said Company shall pay to the said Commissioners for the said Town the sum of dollars and cents, as the settled compensation and damages for taking and using the public highway in said town for a Plank Road, to wit: [describe the road particularly.]

Dated &c,

A. B. Supervisor,
C. D. Commissioner,
E. F. Secy. S. P. R. Co.

(No. 92.)

Affidavit of Nonresidence.—§ 29.

Orleans County, ss.

H. S. of Holley being duly sworn saith that C. D. to whom was awarded the sum of $ by a jury summoned to ascertain the compensation and damages due to him for certain lands intended to be taken and used by the Medina Plank Road Company for the construction of a Plank Road, and whose verdict thereupon was rendered on the day of 18 , is a non resident of this state [or, after diligent search cannot be found in this state, so that the money can be paid to him,) according to the best information and belief of this deponent.

Sworn &c. H. S.

(No. 93.)

Order thereupon.—§ 29.

Satisfactory proof having been made to me that C. D. is a non-resident of this state, (or cannot be found in the state after diligent search), it is ordered that the sum of $, being the amount of verdict of the jury for the compensation and damages awarded him as owner of certain lands to be taken by the Albion Plank Road Company and used for a Plank Road, be paid to the Treasurer of the county of Orleans, (where the lands lie) for the use of such owner, the said C. D. and that notice of such payment be given by publishing the same once in each week, for six successive weeks in a newspaper published in said county of Orleans. H. R. C.

Orleans Co. Judge.

(No. 94.)

Affidavit of Payment and Publication.—§ 29

ORLEANS COUNTY, ss.

K. L. of Albion, being duly sworn, saith that the sum of $ awarded to C. D., by a jury as compensation and damages to be paid him by the Albion Plank Road Company for taking and using his lands for the construction of a Plank Road, has been paid to the Treasurer of the said county as required by order of the Orleans county Judge, and that a notice thereof, of which the annexed is a copy, (annex the printed notice) has been published six successive weeks in the Orleans Republican as required by law.

Sworn &c. K. L.

(No. 95.)

Order thereon.—§ 29.

Satisfactory proof having been made to me, that the sum of $ has been paid to the Treasurer of the county of Orleans, by the Albion Plank Road Company, for the use of C. D., a nonresident owner of lands, as his compensation and damages awarded him for taking and using certain lands owned by him, in the construction of the Plank Road of said Company, and that notice of such payment has been published as required by law, it is ordered, that the said Company take and hold the lands in respect to which the said damages were awarded, in the same manner and with the same effect as if such payment had been made to the owner personally.

H. R. C.

Dated &c. Orleans Co. Judge.

(No. 96.)

Notice of Payment to Treasurer.—§ 29

Notice is hereby given, that the sum of $ has been paid to the Treasurer of Orleans county, by the Albion Plank Road Company, for the use of C. D., being the amount awarded to him as compensation and damages for lands intended to be taken and used by said company in the construction of a Plank Road from to

J. K.

Dated &c. Secy. A. R. Co.

(No. 97.)

Bond to be given by Company on application of Owner for a New Trial.—§ 30.

Know all men by these presents, that we, E. W. and V. S., directors of the Waterloo Plank Road Company as principals, and O. H. as surety, all of Waterloo aforesaid, in the county of Seneca, are held and firmly bound unto E. D. of the town of Seneca Falls, in said county, in the sum of five hundred dollars, (or any proper sum) to be paid to the said E. D., or to his certain attorney, executors, administrators or assigns, for which payment well and truly to be made, we bind ourselves, and our heirs, executors, and administrators, jointly and severally, firmly by these presents. Sealed with our seals, and dated the day of 18 .

The condition of this obligation is such that whereas there has been awarded to the said E. D., the obligee above named, the sum dollars, as compensation and damages for certain lands, intended to be taken and used by the Waterloo Plank Road Company, and the said E. D. having applied for a new trial in the matter, and the said sum having been been deposited as required by law; now therefore, if the said company shall pay the said E. D. any sum of money which he may be entitled to receive from said company, in respect to the land in question by reason of any verdict, or judgment ment of any court for such compensation, damages, costs, and expenses, without fraud or other delay, then this obligation to be void, otherwise to remain in force.

Signed, sealed and delivered in presence of

E. W. [L. S.]
V. S. [L. S.]
O. H. [L. S.]

(No. 98.)

Certificates of Inspectors.—§ 34.

The undersigned Inspectors of plank roads and turnpike roads in and for the county of Niagara, do certify that in pursuance of an

"Act to provide for the incorporation of companies to construct Plank Roads, and of companies to construct Turnpike Roads," passed May 7, 1847, we have inspected the plank road lately constructed by the Lewiston Plank Road Company in said county, and are satisfied that the said road is made and completed according to the true intent and meaning of the said act.

Dated, &c. M. S., N. T., } *Inspectors.*

(No. 99.)

Notice of Commissioners of highways.—§ 37.

The Lockport Plank Road Company will take notice, that we intend to apply to the Niagara County Court, at the next term thereof, to be holden at the court house in the village of Lockport, in and for said county, for an order to alter or change the location of the toll gate now on the road of said company, and situated near the west line of lands of U. K. [Describe the situation of the gate.]

Dated, &c. A. M., L. H., } *Com'rs of Highways of Lewiston.*

To the Lockport Plank Road Company.
[To be served at the office of the company.]

(No. 100.)

Notice of Election of Directors.—§ 38.

Notice is hereby given, that an election of [not less than five nor more than nine] Directors for the Tuscarora Plank Road Company will be held at the office of said company, on the day of , 185 , at 10 o'clock, A. M.

Dated, &c. T. M., *Sec'y.*

(No. 101.)

Notice to pay Stock.—§ 39.

Notice is hereby given, that the stockholders of the Le Roy Plank Road Company are required to pay to the Treasurer of said company, at their office, one fifth part of the stock by them severally subscribed, on or before the day of , 185 , under penalty of forfeiture of their stock and all previous payments thereon.

By order of the Directors. A. L., *Sec'y.*

Dated, &c.

(No. 102.)

Annual Report of Directors.—§ 41.

To the Hon. the Secretary of State of New York :

The undersigned, Directors of the Batavia Plank Road Company, in pursuance of the statute in such case made and provided, beg leave to report: That the road of the said company cost the sum of $;

That the amount of capital stock of said company is $;

Of which sum is already paid the snm of $;

That the amount of all money expended by the company is $;

That the amount of capital stock expended is $;

That the whole amount of tolls or earnings expended on said road is $;

That the amount received during the last year for tolls is $;

That the amount received from all other sources during the year past, to wit: by , and by , &c., &c., is $;

That the amount of dividends made by the company is $;

That the amount set apart for a reparation fund is $;

And that the amount of indebtedness of the said company, which indebtedness accrued for is $;

A. B., } *Directors.*
C. D., }

Dated, &c.

Affidavit to be attached to the Report.—§ 41.

GENSEE COUNTY, ss.

A. B., and C. D., who made and signed the within report, being severally duly sworn, say, that the facts stated therein are true.

A. B.
C. D.

Sworn, &c.,

(No. 103.)

Notice where the Office shall be kept.—§ 42.

Notice is hereby given, that the store of E. L , Esq., in the village of Alexander, is designated as the office of the Alexander Plank Road Company, (or any place in the county where the road is,) until further notice according to law.

A. B., *Sec'y.*

Book containing the names of Stockholders, alphabetically arranged.—§ 43.

Names.	Residence.	No. of Shares.	Time when became holders.
A.			
Allen, George W.	Byron.	Forty.	January 1, 1847.
Adgate, William	Batavia.	Thirty.	October 10, 1847.

[No. 104.]

Certificate of Inspectors presiding at Elections.—§ 9, *Title* 1, *Chap.*, 18, *Revised Statutes.*

The undersigned having presided at an election of Directors for the Cold Spring Plank Road Company, held at in the town of , and county of Putnam, on the day of , 185 , and estimated the votes received at such election, do certify, that there were on that occasion votes cast; of which were for A. B , were for B. D , were for E. F , were for G. H , (naming all the candidates,) and that such election resulted in the choice of A. B , B. D , E. F , G. H , and I. K , as Directors for said company.

Dated, &c.

S. M.
T. U. } *Inspectors.*
R. S.

(No. 105.)

Release of lands under § 1 *of Act November* 24, 1847.

In consideration of the sum of one dollar, to me in hand paid by the Charlotte Plank Road Company, I, A. B., the owner of certain lands desired by said company for the construction of their plank road, to wit: [describe the lands accurately,] do hereby assign, sell and release to said company, the said described land for the proper use and purpose of said company, under the "Plank Road Act," so called, passed May 7, 1847, and "an act in relation to Plank Road and turnpike Companies," passed, Nov. 24, 1847, so long as the said shall, by law, exist.

Signed, sealed and delivered, this day of , 185 , in presence of } A. B. [L. S.]

[No. 106.]

Agreement by town Officers.—§ 1, *Act Nov.* 24, 1847.

The Subscribers, the Supervisors and Commissioners of highways of the town of Greece in the county of Monroe hereby agree to and with the Charlotte Plank Road Company, in consideration of the public benefit to result therefrom, and in pursuance of "An act in relation to Plank Road and Turnpike Road Companies," passed November 24, 1847, to grant and allow to said company, and do hereby grant and allow to said company the right to take and use any part of any public highway in said town necessary for the construction of their road, according to the provisions of the law in such case made and provided.

D. G., *Supervisor.*

F. M.,
S. O., } *Commissioners.*

Dated, &c.

INDEX.

A

B

C

GENERAL LAWS
OF THE STATE OF NEW YORK.

RAIL-ROAD LAWS,

PASSED APRIL 2, 1850, AND AMENDMENT OF 1851.

COMMON SCHOOL LAW,

OF 1851.

ASSESSORS' LAW,

PASSED APRIL 15, 1851.

ROCHESTER:
PUBLISHED BY DAVID HOYT.
1851.

Chap. 140.

AN ACT to authorize the formation of railroad corporations, and to regulate the same.

Passed April 2, 1850, "three fifths being present.

The People of the State of New-York, represented in Senate and Assembly, do enact as follows:

§ 1. Any number of persons not less than twenty-five, may form a company for the purpose of constructing, maintaining and operating a railroad for public use in the conveyance of persons and property, or for the purpose of maintaining and operating any unincorporated railroad already constructed, for the like public use; and for that purpose may make and sign articles of association, in which shall be stated the name of the company; the number of years the same is to continue; the places from and to which the road is to be constructed, or maintained and operated; the length of such road as near as may be, and the name of each county in this state through or into which it is made, or intended to be made; the amount of the capital stock of the company, which shall not be less than ten thousand dollars for every mile of road constructed, or proposed to be constructed, and the number of shares of which said capital stock shall consist, and the names and places of residence of thirteen directors of the company, who shall manage its affairs for the first year, and until others are chosen in their places. Each subscriber to such articles of association shall subcribe thereto his name, place of residence, and the number of shares of stock he agrees to take in said company. On compliance with the provisions of the next section, such articles of association may be filed in the office of the secretary of state, who shall endorse thereon the day they are filed, and record the same in a book to be provided by him for that purpose; and thereupon the persons who have so subscribed such articles of association, and all persons who shall become stockholders in such company, shall be a corporation by the name specified in such articles of association, and shall possess the powers and privileges granted to the corporations, and be subject to the provisions contained in title three of chapter eighteen of the first part of the Revised Statutes, except the provisions contained in the seventh section of the said title.

§ 2. Such articles of association shall not be filed and recorded in the office of the secretary of state, until at least one thousand dollars of stock for every mile of railroad proposed to be made is subscribed thereto, and ten per cent paid thereon in good faith, and in cash, to the directors named in said articles of association; nor until there is endorsed thereon, or annexed thereto, an affidavit made by at least three of the directors named in said articles, that the amount of stock required by this section has been in good faith subscribed, and ten pei cent paid in cash thereon as aforesaid, and that it is intended in good faith to construct or to maintain and operate the road mentioned in such articles of association; which affidavit shall be recorded with the articles of association as aforesaid.

§ 3. A co|y of any articles of association filed and recorded in pursuance with this act, or of the record thereof, with a copy of the affidavit aforesaid endorsed thereon or annexed thereto, and certified to be a copy by the secretary of this state, or his deputy, shall be presumptive evidence of the incorporation of such company, and of the facts therein stated.

§ 4. When such articles of association and affidavit are filed and recorded in the office of the secretary of state, the directors named in said articlcs of association may, in case the whole of the capital stock is not before subscribed, open books of subscription to fill up the capital stock of the company, in such places and after giving such notice as they may deem expedient, and may continue to receive subscriptions until the whole capital stock is subscribed. At the time of subscribing. every subscriber shall pay to the directors ten per cent on the amount subscribed by him, in money; and no subscription shall be received or taken without such payment.

§ 5. There shall be a board of thirteen directors of every corporation formed under this act, to manage its affairs. Said directors shall be chosen annually, by a majority of the votes of the stockholders voting at such election, in such manner as may be prescribed in the by-laws of the corporation, and they may and shall continue to be directors until others are elected in their places. In the election of the directors, each stockholder shall be entitled to one vote for each share of stock held by him. Vacancies in the board of directors shall be filled in such manner as shall be prescribed by the by-laws of the corporation. Every corporation formed under this act shall be subject to the regulations concerning the election of directors of moneyed corporations, contained in article second of the second title of the eighteenth chapter of the first part of the Revised Statutes. The inspectors of the first elections of directors shall be appointed by the board of directors named in the articles of association. No person shall be a director, unless he shall be a stockholder, owning stock absolutely in his own right, and qualified to vote for directors at the election at which he shall be chosen. At every election of directors, the books and papers of such company shall be exhibited to the meeting, provided a majority of the stockholders present shall require it.

§ 6. The directors shall appoint one of their number president: they may also appoint a treasurer and secretary, and such other officers and agents as shall be prescribed by the by-laws.

§ 7. The directors may require the subscribers to the capital stock of the company to pay the amount by them respectively subscribed, in such manner and in such instalments as they may deem proper. If any stockholder shall neglect to pay any instalment as required by a resolution of the board of directors, the said board shall be authorised to declare his stock, and all previous payments thereon, forfeited for the use of the company; but they shall not declare it so forfeited, until they shall have caused a notice in writing to be served on him personally, or by depositing the same in the post-office, properly directed to him at the post-office nearest his usual place of residence, stating that he is required to make such payment at the time and place specified in said notice; and that if he fails to make the same, his stock, and all previous payments thereon, will be forfeited for the use of the company; which notice shall be served as aforsaid, at least sixty days previous to the day on which such payment is required to be made.

§ 8. The stock of every company formed under this act shall be deemed personal estate, and shall be transferable in the manner prescribed by the by-laws of the company, but no shares shall be transferable until all previous calls thereon shall have been fully paid in; and it shall not be lawful for such company to use any of its funds in the purchase of any stock in its own, or in any other corporation.

§ 9. In case any capital stock of any company formed under this act, is found to be sufficient for constructing and operating its road, such campany may, with the concurrence of two thirds in amount of all its stockholders, increase its capital stock from time to time, to any amount required for the purposes aforesaid. Such increase must be sanctioned by a vote in person, or by proxy, of two-thirds in amount of all the stockholders of the company, at a meeting of such stockholders, called by the directors of the company for that purpose, by a notice in writing to each stockholder to be served on him personally, or by depositing the same, properly folded and directed to him, at the post-office nearest his usual place of residence, in the post office, at least twenty days prior to such meeting. Such notice must state the time and place of the meeting, and its object, and the amount to which it is proposed to increase the capital stock. The proceedings of such meeting must be entered on the minutes of the proceedings of the company, and thereupon the capital stock of the company may be increased to the amount sanctioned by a vote of two-thirds in amount of all the stockholders of the company as aforesaid.

§ 10. Each stockholder of any company formed under this act, shall be individually liable to the creditor of such company, to an amount equal to the amount unpaid on the stock held by him, for all the debts and liabilities of such company, until the whole amount of the capital stock so held by him shall have been paid to the company ; and all the stockholders of every such company shall be jointly and severally liable for all the debts due or owning to any of its laborers and servants, for services performed for such corporation ; but shall not be liable to an action therefore, before an execution shall be returned unsatisfied in whole, or in part, against the corporation; and then the amount due on such execution shall be the amount recoverable, with costs, against such stockholders.

§ 11. No person holding stock in any such company, as executor, administrator, guardian or trustee, and no person holding such stock as collateral security, shall be personally subject to any liability as stockholders of such company ; but the person pledging such stock shall be considered as holding the same, and shall be liable as a stockholder accordingly ; and the estates and funds in the hands of such executor, administrator, guardian or trustee, shall be liable in like manner, and to the same extent as the testator, or intestate, or the ward or person interested in such trust fund would have been, if he had been living and competent to act, and held the same stock in his own name.

§ 12. As often as any contractor for the construction of any part of a railroad, which is in progress of construction, shall be indebted to any laborer, for thirty or any less number of days labor performed in constructing said road, such laborer may give notice of such indebtedness to said company in the manner herein provided; and said company shall thereupon become liable to pay such laborer the amount so due him for such labor, and an action may be maintained against said company therefor. Such notice shall be given by said laborer to said company, within twenty days after the performance of the number of days labor for which the claim is made. Such notice shall be in writing, and shall state the amount and number of days labor, and the time when the same was performed, for which the claim is made, and the name of the contractor from whom due, and shall be signed by such laborer, or his attorney; and shall be served on an engineer, agent or superintendent employed by said company, having charge of the section of the road on which such labor was performed, personally, or by leaving the same at the office or usual place of business of such engineer, agent or superintendent, with some person of suitable age. But no action shall be maintained against any company under the provisions of this section, unless the same is commenced within thirty days after notice is given to the company by such laborer as above provided.

§ 13. In case any company formed under this act is unable to agree for the purchase of any real estate required for the purposes of its incorporation, it shall have the right to acquire title to the same, in the manner and by the special proceedings prescribed in this act.

§ 14. For the purpose of acquiring such title, the said company may present a petition, praying for the appointment of commissioners of appraisal, to the supreme court, at any general or special term thereof held in the district in which the real estate described in the petition is situated. Such petition shall be signed and verified according to the rules and practice of such court. It must contain a description of the real estate which the company seek to acquire; and it must, in effect state that the company is duly incorporated, and that it is the intention of the company, in good faith, to construct and finish a railroad from and to the places named for that purpose in its articles of association; that the whole capital stock of the company has been in good faith subscribed as required by this act; that the company has surveyed the line or route of its proposed road, and made a map or survey thereof, by which route or line is designated, and that they have located their said road according to such survey, and filed certificates of such location, signed by a majority of the directors of the company, in the clerks' office of the several counties through or into which the said road is to be constructed; that the land described in the petition is required for the purpose of constructing or operating the proposed road; and that the company has not been able to acquire title thereto, and the reason of such inability. The petition must also state the names and places of residence of the parties, so far as the same can by reasonable diligence be ascertained, who own or have, or claim to own or have estates or interests in the said real estate; and if any such persons are infants, their ages as near as may be, must be stated; and if any of such persons are idiots, or persons of unsound mind; or are unknown, that fact must be stated, together with such other allegations and statements of liens or incumbrances on said real estate as the company may see fit to make. A copy of such petition, with a notice of the time and place the same will be presented to the supreme court, must be served on all persons whose interests are to be affected by the proceedings, at least ten days prior to the presentation of the same to the said court.

1. If the person on whom such service is to be made, resides in this state, and is not an infant, idiot or person of unsound mind, service of a copy of such petition and notice must be made on him or his agent or attorney, authorised to contract for the sale of the real estate described in the petition, personally, or by leaving the same at the usual place of residence of the person on whom service must be made as aforesaid, with some person of suitable age.

2. If the person on whom such service is to be made resides out of the state, and has an agent residing in this state, authorised to contract for the sale of the real estate described in the petition, such service may be made on such agent, or on such person personally out of the state; or it may be made by publishing the notice, stating briefly the object of the application, and giving a description of the land to be taken, in the state paper, and in a paper printed in the county in which the land to be taken is situated, once in each week for one month next previous to the presentation of the petition. And if the residence of such person residing out of this state, but in any of the United States, or any of the British colonies in North America, is known, or can by reasonable diligence be ascertained, the company must, in addition to such publication as aforesaid, deposit a copy of the petition and notice in the post-office, properly folded and directed to such person at the post-office nearest his place of residence, at least thirty days before presenting such petition to the court, and paythe postage chargeable thereon in the United States.

3. If any person on whom such service is to be made is under the age of twenty-one years, and resides in this state, such service shall be made as aforesaid on his general guardian; or if he has no such guardian, then on such infant personally, if the is over the age of fourteen years: and if under that age, then on the person who has the care of, or with whom such infant resides.

4: If the person on whom such service is to be made is an

idiot, or of unsound mind, and resides in this state, such service may be made on the committee of his person or estate; or if has no suchcommitte, then on the person who has the care and charge of such idiot or person of unsound mind

5. If the person on whom such service is to be made is unknown, or his residence is unknown, and can not by reasonable diligence be ascertained, then such service may be made, under the direction of the court, by publishing a notice, stating the time and place the petition will be presented, the object thereof, with a description of the land to be affected by the proceedings, in the state paper, and in the paper printed in the county where the land is situated, once in each week for one month previous to the presentation of such petition.

6. In case any party to be affected by the proceedings is an infant, idiot, or of unsound mind, and has no general guardian or committee, the court shall appoint a special guardian or committee to attend to the interests of such person in the proceedings; but if a general guardian or committee has been appointed for such person in this state, it shall be the duty of such general guardian or committee to attend to the interests of such infant, idiot or person of unsound mind; and the court may require such security to be given by such general or special guardian or committee as it may deem necessary to protect the rights of such infant, idiot or person of unsound mind; and all notices required to beserved in the progress of the proceedings, may be served on such general or special guardian or committee.

7. In all cases not herein otherwise provided for, service of orders, notices, and other papers in the special proceedings authorised by this act, may be made as the supremecourt shall direct.

§ 15. On presenting such petition to the supreme court as aforesaid, with proof of service of a copy thereof and notice as aforesaid, all persons whose estates or interests are to be affected by the proceedings may show cause against granting the prayer of the petition, and may disprove any of the facts alleged in it. The court shall hear the proofs and allegations of the parties, and if no sufficient cause is shown against granting the prayer of the petition, it shall make an order for the appointment of five disinterested and competent persons who reside in the county where the premises to be appraised are situated, commissioners to ascertain and appraise the compensation to be made to the owners or persons interested in the real estate proposed to be taken in such county for the purposes of the company, and to fix the time and place for the first meeting of such commissioners. The parties whose lands are to be appraised, or their attorneys, may, in case they appear, name six such persons, and the company a like numder, provided they do so, and the court shall appoint two of the commissioners from each of the six so named, in case there is no legal objection to such appointment, and the other commissioner shall be appointed by the court in its discretion.

§ 16. The commissioners shall take and subscribe the oath prescribed by the twelfth article of the constitution. Any one of them may issue subpœnas administer oaths to witnesses, and any three of them may adjourn the proceedings before them from time to time, in their discretion. Whenever they meet, except by the appointment of the court or pursuant to adjournment, they shall cause reasonable notice of such meetings to be given to the parties who are to be affected by their proceedings, or their attorney or agent. They shall view the premises described in the petition, and hear the proofs and allegations of the parties, and reduce the testimony, if any is taken by them, to writing; and after the testimony is closed in each case, and without any unnecessary delay, and before proceeding to the examination of any other claim, a majority, of them, all being present and acting, shall ascertain and determine the compensation, which ought justly to be made by the company to the party or parties owing or interested in the real estate appraised by them; and in determining the amount of such compensation they shall not make an allowanceor deduction on account of any real or supposed benefits which the parties in interest may derive from the construction of the proposed railroad. They, or a majority of them, shall also determine and certify what sum ought to be paid to a general or special guardian or committee of an infant, idiot or person of unsound mind, or to an attorney appointed by the court to attend to the interest of any unknown owner or party in interest, not personally served with notice of the proceedings, and who has not appeared, for costs, expenses and counsel fees. They shall make a report to the supreme court, signed by them or a majority of them, of the proceedings before them, with the minutes of the testimony taken by them, if any. Said commissioners shall be entitled to three dollars for their expenses and services for each day they are engaged in the performance of their duties, to be paid by the company.

§ 17. On such report being made by said commissioners, the company shall give notice to the parties or their attorneys to be affected by the proceedings, according to the rules and practice of said court, at a general or special term thereof, for the confirmation of such report; and the court shall thereupon confirm snch report, and shall make an order, con taining a recital of the substance of the proceedings in the matter of the appraisal, and a description of the real estate appraised for which compensation is to be made; and shallalso direct to whom the money is to be paid, or in what bank, and in what manner it shall be deposited by the company.

§ 18. A certified copy of the order so to be made as aforesaid, shall be recorded at full length in the clerk's office of the county in which the land described in it is situated; and thereupon, and on the payment or deposit by the company of the sums to be paid as compensation for the land, and for costs, expenses, and counsel fees as aforesaid, and as directed by said order, the company shall be entitled to enter upon, take possession of, and use the said land for the purposes of its incorporation, during the continuance of its corporate existence, by virtue of this or any other act; and all persons who have been made parties to the proceedings shall be divested and barred of all right, estate and interest in such real estate during the corporate existence of the company as aforesaid. All real estate acquired by any company under and pursuant to the provisions of this act, for the purposes of its incorporation shall be deemed to be acquired for public use. Within twenty days after the confirmation of the report of the commissioners, as provided for in the seventeenth section of this act, either party may appeal, by notice in writing to the other, to the supreme court, from the appraisal and report of the commissioners. Such appeal shall be heardby the supreme court, at any general or special term thereof, on such notice thereof being given, according to the rules and practice of said court. On the hearing of such appeal, the cnurt may direct a new appraisal before the same or new commissioners in its discretion; the second report shall be final and conclusive on all the parties interested. If the amount of the compensation to be made by the company is increased by the second report, the difference shall be a lien on the land appraised, and shall be paid by the company to the parties entitled to the same, or shall be deposited in the bank, as the court shall direct; and if the amount is diminished, the differance shall be refunded to the company by the party to whom the same may have been paid; and judgment therefor may be rendered by the court, on the filing of the second report against the party liable to pay the same. Such appeal shall not affect the possession by such company of the land appraised; and when the same is made by others than the company, it shall not be heard, except on a stipulation of the party appealing not to disturb such possession.

§ 19. If there are adverse and conflicting claimants to the money, or any part of it, to be paid as compensation for the real estate taken, the court may direct the money to be paid into the said court by the company, and may determine who is entitled to the same, and direct to whom the same shall be paid; and may, in its discretion, order a reference to ascertain the facts on which such determination and order are to be made.

§ 20. The court shall appoint some competent attorney to appear for, and protect the rights of any party in interest, who is unknown, or whose residence is unknown, and who has not appeared in the proceedings by an attorney or agent. The court shall also have power at the time to maend any defect or informality in any of the special proceedings authorised by this act, as may be necessary; or to cause new parties to be added, and to direct such further notices to be given, to any party in interest, as it deems proper; and also to appoint other commissioners in place of any who shall die, or refuse, or neglect to serve, or be incapable of serving.

§ 21. If, at any time after an attempt to acquire title by appraisal of damages or otherwise, it shall be found that the title thereby attempted to be acquired is defective, the company may proceed anew to acquire or perfect such a title in the same manner as if no appraisal had been made; and at any stage of such new proceedings, the court may authorise the corporation, if in possession, to continue in possession, and if not in possession, to take possession, and use such real estate during the pendency and until the final conclusion of such new proceedings; and may stay all actions or prceedings against the company on account thereof, on such company paying into court a sufficient sum, or giving security as the court may direct, to pay the compensation therefor when finally ascertained; and in every such case, the party interested in such real estate may conduct the proceedings to a conclusion, if the company delays or omits to prosecute the same.

§ 22. Every company formed under this act, before constructing any part of their road into or through any county named in their articles of association, shall make a map and profile of the route intended to be adopted by such company in such county, which shall be certified by the president and engineer o the company, or a majority of the directors, and filed in the office of the clerk of the county in which the road is to be made. The company shall give written notice to all actual occupants of the land over which the route of the road is so designated, and which has not been purchased by or given to the company, of the route so designated. Any party feeling aggrieved by the proposed location, may, within fifteen days after receiving written notice as aforesaid, apply to a justice of the supreme court, out of court by petition, duly verified, setting forth his objections to the route designated; and the said justice may, if he considers sufficient eause therefor to exist, appoint three disinterested persons, one of whom must be a practical engineer, commissioners to examine the proposed route, and after hearing the parties, to affirm or alter the same, as may be consistent with the just rights of all parties and the public; but no alteration of the route shall be made, except by the concurrence of the commissioner who is a practical civil engineer. The determination of the commissioners shall, within thirty days after their appointment be made and certified by them, and the certificate filed in the office of the county clerk. Said commissioners shall each be entitled to three dollars per day for their expenses and services, to be paid yb the person who applied for their appointment; and if the proposed route of the road is altered or changed by the commissioners, the company shall refund to amount so paid.

§ 23. The directors of every company formed under this act may, by a vote of two-thirds of their whole number, at any time alter or change the route or any part of the route of their road, if it shall appear to them that the line can be improved thereby; and they shall make and file in the clerk's office of the proper county, a survey, map and certificate of such alteration or change; and shall have the same right and power to acquire title to any lands required for the purposes of the company, in such altered or changed route, as if the road had been located there in the first instance; and no such alteration shall be made in any city or village, after the road shall have been constructed, unless the same is sanctioned by a vote of two-thirds of the common council of said city or trustees of said village; and in case of any alteration made in the route of any railroad after the company has commenced grading, compensation shall be made to all persons for injury so done to any lands that may have been donated to the company. All the provisions of this act relative to the first location, and to acquiring title to land, shall apply to every such new or altered portion of the route.

§ 24. Whenever the track of a railroad constructed by a company formed under this act shall cross a railroad, a highway, turnpike or plank road, such highway, turnpike or plank road may be carried under or over the track, as may be found most expedient; and in cases where an embankment or cutting shall make a change in the line of such highway, turnpike or plank road desirable, with a view to a more easy ascent or descent, the said company may take such additional lands for the construction of such road, highway, turnpike or plank road on such new line as may be deemed requisite by the directors. Unless the lands so taken shall be purchased for the purposes aforesaid, compensation therefor shall be ascertained in the manner prescribed in this act for accquiring title to real estate, and duly made by said corporation to the owners and persons interested in such lands. The same when so taken, shall become part of such intersecting highway, turnpike or plank road, in such manner and by such tenure as the adjacent parts of the same highway, turnpike or plank road, may be held for highway purposes.

§ 25. The commissioners of the land office shall have power to grant to any railroad company formed under this act, any land belonging to the people of this state, which may be required for the purposes of their road, on such terms as may be agreed on by them; or such company may acquire title thereto by appraisal, as in the case of lands owned by individuals; and if any land belonging to the county or town is required by any company for the purposes of the road, the county or town officers having the charge of such land may grant such land to such company, for such conpensation as may be agreed upon.

§ 26. In case any title or interest in real estate required by any company formed under this act, for the purpose of its incorporation, shall be vested in any trustee not authorised to sell, release and convey the same, or in any infant, idiot or person of unsound mind, the supreme court shall have power, by a summary proceeding on petition, to authorise and empower such trustee, or the general guardian or committee of such infant, idiot or person of unsound mind, to sell and convey the same to such company, for the purposes of its incorporation, on such terms as may be just; and in case any such infant, idiot or person of unsound mind, has no general guardian or committee, the said court may appoint a special guardian or committee for the purpose of making such sale, release or conveyance, and may require such security from such general or special guardian or committee as said court may deem proper. But before any conveyance or release authorised by this section shall be executed, the terms on which the same is to be executed shall be reported to the court, on oath; and if the court is satisfied that such terms are just to the party interested in such real estate, the court shall confirm the report, and direct the proper conveyance or release to be executed, which shall have the same effect as if executed by an owner of said land, having legal power to sell or convey the same.

§ 27. No company formed under this act shall lay down or use in the constuction of their road, any iron rail of less weight than fifty-six pounds to the lineal yard, except for turnouts, sidings and switches.

§ 28. Every corporation formed under this act, shall in addition to the powers conferred on corporations in third title of the eighteenth chapter of the first part of the Revised Statutes, have power.

1. To cause such examination and surveys for its proposed railroad to be made, as may be necessary to the selection of the most advantageous route; and for such purpose, by its officers or agents and servants, to enter upon the lands or waters of any person, but subject to responsibility for all damages which shall be done thereto.

2. To take and hold such voluntary grants of real estate and other property as shall be made to it, to aid in the construction, maintainance and accommodation of its railroad; but the real estate received by voluntary grant shall be held and used for the purposes of such grant only.

3. To purchase, hold and use all such real estate and other property as may be necessary for the construction and maintainance of its railroad, and the stations and other accommodations necessary to accomplish the objects of its incorporation; but nothing herein contained shall be held as repealing, or in any way affecting the act entitled "An act authorizing the construction of railroads upon Indian lands," passed May 12, 1836.

4. To lay out its road not exceeding six rods in width, and to construct the same; and for the purposes of cuttings and embankments, to take as much more land as may be necessary for the proper construction and security of the road, and to cut down any standing trees that may be in danger of falling on the road, making compensation therefor as provided in this act for lands taken for the use of the company.

5. To construct their road across, along, or upon any stream of water, water-course, street, highway, plank road, turnpike, or canal, which the route of its road shall intersect or touch; but the company shall restore the stream or water-course, street, highway, plank road and turnpike thus intersected or touched, to its former state, or to such state as not unnecessarily to have impaired its usefulness. Every company formed under this act, shall be subject to the power vested to the canal commissioners by the seventeenth section of chapter two hundred and seventy six of the session laws of eighteen hundred and thirty-four. Nothing in this act contained shall be construed to authorise the erection of any bridge, or any other obstruction across, in or over any stream or lake navigated by steam or sail boats, at the place where any bridge or other obstructions may be proposed to be placed; nor to authorise the construction of any railroad not already located in, upon or across any streets in any city, without the assent of the corporation of such city.

6. To cross, intersect, join and unite its railroad with any other railroad before constructed, at any point on its route, and upon the grounds of such other railroad company, with the necessary turnouts, sidings and switches, and other conveniences in furtherance of the objects of its connections. And every company whose railroad is or shall be hereafter intersected by any new railroad, shall unite with the owners of such new railroad in forming such intersections and connections, and grant the facilities aforesaid; and if the two corporations cannot agree upon the amount of compensation to be made therefor, or the points and manner of such crossings and connections, the same shall be ascertained and determined by commissioners to be appointed by the court as is provided in this act in respect to acquiring title to real estate.

7. To take and convey persons and property on their rail. road by the power or force of steam or of animals, or by any mechanical power, and to receive compensation therefor.

8. To erect and maintain all necessary and convenient buildings, stations, fixtures and machinery for the accomodation and use of their passengers, freights and business.

9. To regulate the time and manner in which passengers and property shall be transported, and the compensation to be paid therefor,; but such compensation, for any passenger and his ordinary baggage, shall not exceed three cents per mile.

10. From time to time to borrow such sums of money as may be necessary for completing and finishing or operating their railroad, and to issue and dispose of their bonds for any amount so borrowed, and to mortgage their corporate property and franchises to secure the payment of any debt contracted by the company for the purposes aforesaid; and the directors of the company may confer on any holder ot any bond issued for money borrowed as aforesaid, the right to convert the principal due or owing thereon, into stock of said company, at any time not exceeding ten year from the date of the bond, under such regulations as the directors may see fit to adopt.

§ 29. Whenever the railroad of any company formed under this act shall run parallel or nearly parallel to any canal of this state, and within thirty miles of such canal, the company owning such railroad shall pay to the canal fund, on all property transported upon its railroad other than the ordinary baggage of passengers, the same tolls upon that portion of the road running parallel to the canal, that would have been payable to the state if such property other than baggage had been transported on any such canal; and every such company shall make returns, at such times and in such manner as the commissioners of the canal fund shall prescribe, of all the property transported on its railroad, except ordinary baggage of passengers; and the said commissioners are authorised to prescribe the manner in which tolls so payable to the canal fund by such company, shall be collected and paid, and to enforce the collection and payment thereof, and to make such regulations as they shall deem proper for that purpose; and every such company that shall neglect or refuse to comply with any such regulations, shall forfeit to the people of this state the sum of five hundred dollars for every day it shall so neglect or refuse; and in every case of such forfeiture, it shall be the duty of the attorney general to prosecute such company for the penalty, in the name of the people.

§ 30. Every conductor, baggage master, engineer, brakeman, or other servant of any railroad corporation employed in a passenger train, or at stations for passengers, shall wear upon his hat or cap a badge, which shall indicate his office, and the initial letters of the style of the corporation by which he is employed. No conductor or collector without such badge shall be entitled to demand or receive from any passenger any fare or ticket, or to exercise any of the powers of his office; and no officer or servant, without such badge, shall have authority to meddle or interfere with any passenger, his baggage or property.

§ 31. Every railroad corporation formed under this act, shall make an annual report to the state engineer and surveyor of the operations of the year ending on the thirteenth day of September; which report shall be verified by the oaths of the treasurer, or president, and acting superintendent of operations, and be filed in the office of the state engineer and surveyor by the first day of December in each year, and shall state:

1. The amount of capital as by charter;
2. The amount of stock subscribed;
3. The amount paid in as by last report;
4. The total amount now of capital stock paid in;
5. The funded debt by last report;
6. The total amount now of funded debt;
7. The floating debt as by last report;
8. The amount now of floating debt;
9. The total amount now of funded and floating debt;
10. The average rate per annum of interest on funded debt;

Cost of road and equipment.

11. For graduation and masonary by last report;
12. The total amount now expended for the same;
13. The amount for bridges by last report;
14. The total amount now expended for the same.
15. The am't for superstructure, including iron, by last report;

16. Total amount now expended for the same.
17. For passengers and freight stations, building and fixtures, by last report:
18. Total amount now expended for the same.
19. For engine and car houses, machine shops, and machinery and fixtures, by last report;
20. Total amount now expended for the same.
21. For land, land damages and fences, by last report;
22. Total amount now expended for the same.
23. For locomotives & fixtures & snow plows, by last report;
24. Total amount now expended for the same.
25. For passenger and baggage cars, by last report;
26. Total amount now expended for the same.
27. For freight cars, as by last report;
28. Total amount now expended for the same.
29. For engineering and agencies, by last report:
30. Total amount now expended for the same.
31. Total cost of road and equipment.

Characteristics of road.

32. Length of road;
33. Length of road laid;
34. Length of double track, including sidings;
35. Length of branches owned by the company laid;
36. Length of double track on the same.
37. Weight of rail by yard on main track.
38. The number of engine houses and shops; of engines and cars, and their character.
39. It shall also be the duty of each corporation to transmit to the state engineer and surveyor the following maps, profiles and drawing exhibiting the characteristics of their roads; the map to show the length and direction of each straight line, and the length and radius of each curve; also the point of crossing of each town and county line, and the length of line in each town and county accurately determined by measurements to be taken after the completion of the road. The profile to be on the map, and shall show the grade line and surface of ground in the usual method, also the elevation of grades above tides at each change in the inclination thereof. The maps and profile to be made on a scale of five hundred feet to one-tenth of a foot; vertical scale of five hundred feet to one tenth of a foot. For all roads or parts of roads now done, or in operation, the said maps shall be returned on or before the first day of January next; and for all roads now in progress, or which may hereafter be construsted, the said maps and profiles shall be returned within three months after the same or any portion thereof shall be in use.

Doings of the year in transportation, and total miles run.

40, Miles run by passenger trains:

41. Miles run by freight trains;

42. The rate of fare for passengers, charged for the respective classes per mile;

43. Number of passengers carried in cars;

44. Number of miles travelled by passengers;

45. Number of tons of two thousand pounds of freight carried in cars;

46. Number of miles caried, or total movement of freight in miles: all to be accurately compiled from the daily records or evidences of earnings, manifest and way-bills.

47. Average rate of speed adopted by ordinary passenger trains, including stops;

48. Average rate of speed adopted by ordinary passenger trains when in motion.

49. Average rate of speed adopted by express trains including stops;

50. Average rate of speed adopted by express trains when in motion.

51. Average rate of speed adopted by freight trains, including stops;

52. Average rate of speed adopted by freight trains when in motion.

53. Average weight in tons of two thousand pounds of passenger trains, exclusive of passengers and baggage;

54. Average weight in tons of freight trains, exclusive of freight.

55. The amount of freight, specifying the quantity in tons, of the products of the forest, of animals, of vegetable food, other agricultural products, manufactures, merchandise and other articles.

Expenses of maintaining the road or real estate of the corporation.

56. For repairs of road bed and railway, excepting cost of iron, which shall be the cost of labor and materials used during the year; also use and cost of engines engaged in ballasting; also the renewal and repairs of gravel and stone cars, and all items of cost connected with keeping the road in order.

57. For depreciation of way;

58. Length, in feet, of iron used in renewals, with weight and cost;

59. Repairs of buildings;

60. Repairs of fences and gates;

61. Taxes on real estate;

62. Total expenses of maintaining road or real estate for the year;

63. Expenses of machinery or personal property of the corporation;

64. Repairs of engines and tenders;

65. Depreciation of engines and tenders;

66. Repairs of passenger and baggage cars;

67. Depreciation of passenger and baggage cars;

68. Repairs of freight cars;

69, Depreciation of freight cars;

70. Repairs of tools and machinery in shops;

71. Incidental expenses, including fuel, oil, clerks, watchmen about shops;

72. Total expenses of repair of machinery;

73. Office expenses, stationary;

74. Agents and clerks;

75. Labor handling freight, loading and unloading;

76. Porters, watch and switchmen;

77. Wood and water station attendance;

78. Conductors, baggage and brakemen;

79, Enginemen and firemen:

80. Fuel (first cost, and labor preparing for use);

81. Oil and waste for engines and tenders;

82. Oil and waste for freight cars;

83. Oil and waste for baggage and passenger cars;

84. Loss and damage of goods and baggage;

85. Damages for injuries of persons;

86- Damages to property, including damages by fire, cattle killed on road;

87. General superintendence;

88. Contengencies;

89. Total expenses of operating road;

90. The above statements are to be made without reference to the sums actually received or paid during the year. The following statement of the earnings and cash receipts and payments are required:

91. From passengers;

92. From freight;

93. From other sources;

94. The above to be stated without reference to the amount actually collected;

95. Receipts during the year from freight;

96. From passengers;

97. From other sources, specifying what in detail;

98. Payments for transportation expenses;

99. For interest.

100. Devidends on stock, amount and rate per cent.

101. Payments to surplus fund, and the total amount of said fund.

102. The number of persons injured in life and limb, and the cause of the injury, and whether passengers or persons employed.

Whether any such accidents have arisen from carelessness or negligence of any person in the employment of the corporation, and whether such person is retained in the service of the corporation.

103. It shall be the duty of the state engineer and surveyor to arrange the information contained in such reports in a tabular form, and prepare the same, together with the said reports, in a single document, for printing, for the use of the legislature, and report the same to the legislature on the first day of its session in each year.

104. All the items under the heads of expenses of maintaining the road or real estate of the corporation, expenses of machinery or personal property of the corporation, expenses of use of road and machinery or operating the road, shall be carried out under two heads, the one showing the cost of freight transportation, the other the cost of passenger transportation.

105. The provisions of this section shall apply to all existing railroad corporations; and the report of the said existing railroad corporations, made in pursuance of the provisions of this section, shall be deemed to be a full compliance with any existing law or resolution requiring annual reports to be made by such corporation.

§ 32. Any such corporation which shall neglect to make the report as is provided in the preceeding section, shall be liable to a penalty of two hundred and fifty dollars, to be sued for in the name of the people, for their use.

§ 33. The legislature may, when any such railroad shall be opened for use, from time to time, alter or reduce the rate of freight, fare, or other profits upon such road; but the same shall not, without the consent of the corporation, be so reduced as to produce with said profits less than ten per centum per annum on the capital actually expended; nor unless on an examination of the amounts received and expended, to be made by the state engineer and surveyor, and the comptroller, they shall ascertain that the net income derived by the company from all sources for the year then last past shall have exceeded an annual income of ten per cent. upon the capital of the cor-poration actually expended.

§ 34. Any such corporations shall, when applied to by the

postmaster-general, convey the mails of the United States on their road or roads respectively; and in case such corporation shall not agree as to the rate of transporation therefor, and as to the time, rate of speed, manner and condition of carrying the same, it shall be lawful for the governor of this state to appoint three commissioners, who, or a majority of them, after fifteen days' notice in writing of the time and place of meeting to the corporation, shall determine and fix the prices, terms and conditions aforesaid; but such price shall not be less for carrying said mails in the regular passenger trains, than the amount which such corporation would receive as freight on a like weight of merchandise transported in their merchandise trains, and a fair compensation for the post-office car. And in case the postmaster-general shall require the mail to be carried at other hours, or at a higher speed than the passenger trains are run, the corporation shall furnish an extra train for the mail, and be allowed an extra compensation for the expenses, and wear and tear thereof, and for the service to be fixed as aforesaid.

§ 35. If any passenger shall refuse to pay his fare, it shall be lawful for the conductor of the train and the servants of the corporation to put him and his baggage out of the cars, using no unnecessary force, at any usual stopping place, or at any dwelling house, as the conductor shall elect, on stopping the train.

§ 36. Every such corporation shall start and run their cars for the transportation of passengers and property, at regular times, to be fixed by public notice; and shall furnish sufficient accommodation for the transportation of all such passengers and property, as shall within a reasonable time previous thereto being offered for transportation at the place of starting and the junctions of other railroads, and at usual stopping places established for receiving and discharging way passengers and freights for that train; and shall take, transport and discharge such passengers and property at, from and to such places, on the due payment of the freight or fare legally authorized therefor; and shall be liable to the party aggrieved, in an action for damages, for any neglect or refusal in the premises.

§ 37. A check shall be affixed to every parcel of baggage, when taken for transportation by the agent or servant of such corporation, if there is a handle, loop or fixture, so that the same can be attached upon the parcel of baggage so offered for transportation, and a duplicate thereof given to the passenger or person delivering the same on his behalf; and if such check be refused on demand, the corporation shall pay to such passenger the sum of ten dollars, to be recovered in a civil action; and further, no fare or toll shall be collected or received from such passenger, and if such passenger shall have paid his fare, the same shall be refunded by the conductor in charge of the train; and on producing said check, if his baggage shall not be delivered to him, he may himself be a witness in any suit brought by him, to prove the contents and value of said baggage.

§ 38. In forming a passenger train, baggage, freight, merchandise, or lumber cars shall not be placed in rear of the passenger cars; and if they or any of them shall be so placed the officer or agent who so directed, or knowingly suffered such arrangement, and the conductor of the train, shall be deemed guilty of a misdemeanor, and be punished accordingly.

§ 39. A bell shall be placed on each locomotive engine, and be rung at the distance of at least eighty rods from the place where the railroad shall cross any traveled public road or street, and be kept ringing until it shall have crossed such road or street; or a steam whistle shall be attached to each locomotive engine, and be sounded at least eighty rods from the place where the railroad shall cross any such road or street, except in cities, and be sounded at intervals until it shall have crossed such road or street, under a penalty of twenty dollars for every neglect of the provisions of this section, to be paid by the corporation owning the railroad, to be sued for by the district attorney of the county, within ten days after such penalty was incurred; one-half thereof to go to the informer, and the other half to the county; and said corporation shall also be liable for all damages which shall be sustained by any person by reason of such neglect, one-half of which penalty shall be chargeable to, and collected by the company, of the engineer having charge of the train, where the omission of duty consists in not sounding the whistle or ringing the bell.

§ 40. Every such corporation shall cause boards to be placed, well supported by posts or otherwise, and constantly maintained across each traveled public road or street where the same is crossed by the railroad, on the same level. Said boards shall be elevated so as not to obstruct the travel, and to be easily seen by travelers; and on each side of such boards shall be painted in capital letters, of at least the size of nine inches each, the words, "Railroad crossing, look out for the cars." But this section shall not apply to streets in cities or villages, unless the corporation shall be required to put up such boards by the officers having charge of such streets.

§ 41. If any person shall, while in charge of a locomotive

engine running upon the railroad of any such corporation, or while acting as the conductor of a car or train of cars on any such railroad, be intoxicated, he shall be deemed guilty of a misdemeanor.

§ 41. If any person or persons shall wilfully do, or cause to be, done, any act or acts whatever, whereby any building construction or work of any railroad corporation, or any engine, machine or structure, or any matter or thing appertaining to the same, shall be stopped, obstructed, impaired, weakened, injured or destroyed, the person or persons so offending shall be guilty of a misdemeanor, and shall forfeit and pay to the said corporation treble the amount of damages sustained by means of such offence.

§ 43. All penalties imposed by this act may be sued for in the name of the people of the state of New-York; and if such penalty e for a sum not exceeding one hundred dollars, then such suib may be brought before a justice of the peace, and may be tcommenced by serving a summons on any director of such company.

§ 44. Every corporation formed under this act, shall erect and maintain fences on the sides of their road, of the height and strength of a division fence required by law, with openings or gates or bars therein, and farm crossings of the road for the use of the proprietors of lands adjoining such railroad; and also construct and maintain cattle-guards at all road crossings, suitable and sufficient to prevent cattle and animals from getting on to the railroad. Until such fences and cattle-guards shall be duly made, the corporation and its agents shall be liable for all damages which shall be done by their agents or engines, to cattle, horses, or other animals thereon; and after such fences and guards shall be duly made and maintained, the corporation shall not be liable for any such damages, unless negligently or wilfully done; and if any person shall ride, lead or drive any horse or other animal upon such road, and within such fences and guards, other than at farm crossings, without the consent of the corporation, he shall for every such offence forfeit a sum not exceeding ten dollars, and shall also pay all damages which shall be sustained thereby to the party aggrieved. It shall not be lawful for any person, other than those connected with or employed upon the railroad, to walk along the track or tracks of any railroad, except where the same shall be laid along public roads or streets.

§ 45. Every corporation shall, within a reasonable time after their road shall be constructed, cause to be made:

A map and profile thereof, and of the land taken or obtained for the use thereof, and file the same in the office of the state engineer and surveyor; and also like maps of the parts thereof located in different counties, and file the same in the offices for recording deeds, in the county in which such parts of said road shall be. Every such map shall be drawn on a scale, and on paper, to be designated by the state engineer and surveyor, and certified and signed by the president or engineer of such corporation.

§ 46. In case any passenger on any railroad shall be injured while on the platform of a car, or on any baggage, wood, or freight car, in violation of the printed regulations of the company posted up at the time in a conspicuous place inside of its passenger cars then in the train, such company shall not be liable for the injury; provided, said company at the time furnished room inside its passenger cars sufficient for the proper accommodation of the passengers.

§ 47. If any corporation formed under this act shall not, within two years after its articles of association are filed and recorded in the office of the secretary of state, begin the construction of its road, and expend thereon ten per cent on the amount of its capital, or shall not finish the road and put it in operation in five years from the time of filing its articles of association as aforesaid, its corporate existence and powers shall cease.

§ 48. The legislature may at any time annul or dissolve any corporation formed under this act; but such dissolution shall not take away or impair any remedy given against any such corporation, its stockholders or officers for any liability which shall have been previously incurred.

§ 49. All existing railroad corporations within this state shall respectively have and possess all the powers and privileges contained in this act; and they shall be subject to all the duties, liabilities and provisions not inconsistent with the provisions of their charter, contained in sections nine, thirteen, fourteen, fifteen, sixteen, seventeen, eighteen, nineteen, twenty, twenty-one, twenty-three, twenty-four, twenty-five, twenty-six, twenty-seven, twenty-eight, (except subdivision nine,) thirty, thirty-one, thirty-two, thirty-three, thirty-four, thirty-five, thirty-six, thirty-seven, thirty-eight, thirty-nine, forty, forty-one, forty-two, forty-three, forty-four, forty-five, forty-six, of this act.

§ 50. The act entitled, "An act to authorize the formation of railroad corporations, passed March 26, 1848, and the acts amending the same are hereby repealed; but all railroad companies formed under said act are hereby continued in existence, in the same manner as if said acts were not repealed and such companies shall be subject to all the provisions, and shall have the same powers, rights and privileges, and be subject to the same duties, as if they had been incorporated under this act; and the time limited by said act, for the expenditure of ten per cent of their capital stock, is hereby extended two years from the passage of this act; and the time limited in said section of said law for their completion, is hereby extended to five years from the passage of this act; and also the time for completing any railroad organized previous to March 27, 1848, whose road was under contract prior to February 1, 1850, to be completed within the time prescribed by its charter, is here extended for one year.

§ 51. Nothing in this act contained shall authorize or permit the New-York and Erie Railroad Company to abandon the use of their road in the county of Rockland, east of Suffern's depot.

§ 52. This act shall take effect immediately.

AN ACT to establish free schools throughout the state.

The People of the State of New York, represented in Senate and Assembly, do enact as follows:

SECTION 1. Common schools in the several school districts in this state shall be free to all persons residing in the district over five and under twenty-one years of age, as hereinafter provided. Persons not resident of a district may be admitted into the schools kept therein with the approbation, in writing, of the trustees thereof, or a majority of them.

§ 2. There shall hereafter be raised by tax, in each and every year, upon the real and personal estate within this state, the sum of eight hundred thousand dollars, which shall be levied, assessed and collected in the mode prescribed by chapter thirteen, part first of the revised statutes, relating to the assessment and collection of taxes, and when collected shall be paid over to the respective county treasurers, subject to the order of the State Superintendent of common schools.

§ 3. The State Superintendent of common schools shall ascertain the portion of said sum of eight hundred thousand dollars to be assessed and collected in each of the several counties of this State, by dividing the said sum among the several counties, according to the valuation of real and personal estate therein, as it shall appear by the assessment of the year next preceding the one in which said sum is to be raised, and shall certify to the clerk of each county, before the tenth day of July in each year, the amount to be raised by tax in such county; and it shall be the duty of the several county clerks of this State to deliver to the Board of Supervisors of their respective counties, a copy of such certtficate on the first day of their annual session, and the Board of Supervisors of each county shall assess such amount upon the real and personal estate of such county, in the manner provided by law for the assessment and collection of taxes.

§ 4. The State Superintendent of common schools shall, on or before the first day of January in every year, apportion and divide, or cause to be apportioned and divided, one-third of the sum so raised by general tax, and one-third of all other monies appropriated to the support of common schools, among the several school districts, parts of districts, and separate neighborhoods in this State, from which reports shall have been received in accordance with law, in the following manner, viz: to each separate neighborhood belonging to a school district in some adjoining State there shall be apportioned and paid a sum of money equal to thirty-three cents for each child in such neighborhood (between the ages of four and twenty-one; but the sum so to be apportioned and paid to any such neighborhood,) shall in no case exceed the sum of twenty-four dollars, and the remainder of such one-third shall be apportioned and divided equally among the several districts; and the State Superintendent of common schools shall, by proper regulations aud instructions to be prescribed by him, provide for the payment of such monies to the trustees of such neighborhoods and school districts.

§ 5. It shall be the duty of the State Superintendent of common schools, on or before the first day of January, in every year, to apportion and divide the remaining two-thirds of the said amount of eight hundred thousand dollars, together with the remaining two-thirds of all other monies appropriated by the State for the support of common schools among the several counties, cities and towns of the State, in the mode now prescribed by law for the division and apportionment of the income of the common school fund; and the share of the several towns and wards so apportioned and divided, shall be paid over, on and after the first Tuesday of February, in each year, to the several town Superintendents of common schools, and ward or city officers, entitled by law to receive the same, and shall be apportioned by them among the several school districts and parts of districts in their several towns and wards, according to the number of children between the ages of four and twenty-one years, residing in said districts and parts of districts, as the same shall have appeared from the last annual report of the trustees; but no monies shall be apportioned and paid to any district or part of a district unless it shall appear from the last annual report of the trustees that a school has been kept therein for at least six months during the year, ending with the date of such report by a duly qualified teacher, unless by special permission of the State Superintendent of common schools; excepting, also, that the first apportionment of money under the act shall be made to all school districts which were entitled to an apportionment of public money in the year eighteen hundred and forty-nine.

§ 6. Any balance required to be raised in any school district for the payment of teachers' wages, beyond the amount apportioned to such district by the previous provisions of this act, and other public monies belonging to the district applicable to the payment of teachers' wages, shall be raised by rate bill to be made out by the trustees against those sending to school, in proportion to the number of days and of children sent, to be ascertained by the teachers' list, and in making out such rate bill it shall be the duty of the trustees to exempt, either wholly or in part, as they may deem expedient, such indigent inhabitants as may, in their judgment, be entitled to such exemption, and the amount of such exemption shall be added to the first tax list thereafter to be made out by the trustees for district purposes, or shall be separately levied by them, as they shall deem most expedient.

§ 7. The same property which is exempt by section twenty-two, of article two, title five, chapter six, part three of the revised statutes from levy and sale under execution, shall be exempt from levy and sale under any warrant to collect any rate bill for wages of teachers of common schools.

§ 8. Nothing in this act shall be so construed as to repeal or alter the provisions of any special act relating to schools in any of the incorporated cities or villages of this State, except so far as they are inconsistent with the provisions contained in the first, second, third and fourth sections of this act.

§ 9. Chapter one hundred and forty of the session laws of one thousand eight hundred and forty-nine, entitled "An act establishing free schools throughout the State," and chapter four hundred and four of the session laws of one thousand eight hundred and forty nine, entitled "An act to amend an act entitled an act establishing free schools throughout the State," and sections sixteen, seventeen and eighteen of the revised statutes relating to common schools, requiring the several Boards of Supervisors to raise by tax, on each of the towns of their respective counties, a sum equal to the school monies appropriated to such towns, and providing for its collection and payment, and all other provisions of law incompatible with the provisions of this act, are hereby repealed.

§ 10. The State Superintendent of common schools shall cause to be prepared, published and distributed among the several school districts and school officers of the State a copy of the several acts now in force relating to common schools, with such instructions, digest and expositions as he may deem expedient; and the expenses incurred by him therefor shall be audited by the Comptroller and paid by the Treasurer.

§ 11. All the monies received or appropriated by the provisions of this act shall be applied to the payment of teachers' wages exclusively.

§ 12. It shall be the duty of the trustees of the several school districts in this State to make out and transmit to the Superintendent of the town in which their respective school houses shall be located, on or before the first day of September next, a correct statement of the whole number of children residing in their district on the first day of August preceding the date of such report between the ages of four and twenty-one; and such town Superintendent shall embody such statement in a tabular form, and transmit the same to the county clerk in sufficient season for the latter to incorporate the information thus obtained in the annual report required to be made to the State Superintendent of common schools for the present year.

§ 13. It shall also be the duty of the trustees of the several school districts, in their annual reports thereafter to be made, to specify the number of children, between the aforesaid ages, residing in their respective districts on the last day of December in each year, instead of the number of such children between the ages of five and sixteen.

§ 14. This act shall take effect on the first day of May next; but nothing herein contained shall be so construed as to affect provisions already made in the several school districts for the support of schools therein under existing laws for the current year.

An Act to amend the law for the Assessment and collection of Taxes.

The People of the State of New York, represented in Senate and Assemby, d enact as follows:

Section 1. Sec. two, article one, title two, chapter thirteen, part first of the Revised Statutes in relation to the assessment and collection of taxes is hereby amended so as to read as follows :—" Land occupied by a person other than the owner, may be assessed to the owner or occupant, or as non-resident lands."

§ 2. Section five of the same title is hereby amended so as to read as follows: "Every person shall be assessed in the town or ward where he resides when the assessment is made, for all personal estate owned by him, including all personal estate in his possession or under his control as agent, trustee, guardian, executor or administrator, and in no case shall the property so held under either of these trusts be assessed against any other person, and in case any person possessed of such personal estate shall reside during any year in which taxes may be levied, in two or more counties, or wards, his residence for the purpose and within the meaning of this section, shall be deemed and held to be in the county, town or ward in which his principal business shall have been transacted, but the products of any State of the United States, consigned to any agents in any town or ward of this State, for sale or commission, for the benefit of the owner thereof, shall not be assessed to such agent, nor shall such agent of moneyed corporations or capitalists be liable to taxation under this section, for any moneys in their possession or under their control, transmitted to them for investment or otherwise."

§ 3. Sections fifteen, sixteen, twenty-two, twenty-three twenty-four, twenty-five, and twenty-six, of the same title are hereby repealed, and section seventeen of the same title is hereby amended so as to read as follows: All real and personal estate liable to taxation, shall be estimated and assessed by the assessors at its full and true value, as they would apppraise the same in payment of a just debt due from a solvent debtor."

§ 4. Section twenty of the same article is hereby amended as to read as follows: Such notice shall set forth that the assessors have completed their assessment roll, and that a copy thereof is left with one of their number at a place to be specified therein, where the same may be seen and examined by any person interested, until the third Tuesday of August; and that on that day the assessors will meet at a time and place also to be specified in such notice, to review their assessments. On the application of any person conceiving himself aggrieved, it shall be the duty of the said assessors on such day to meet, at the time and place specified, and hear and examine all complaints in relation to such assessments that may be brought before them; and they are empowered, and it shall be their duty, to adjourn from time to time as may be necessary, to determine such complaints; but in the several cities of this State, the notice, as required by this section, may conform to the requirements of the respective laws regulating the time and place and manner for revising the assessments in said cities, in all cases where a different time, place and manner is prescribed by said laws from that mentioned in this act.

§ 5. If the assessors shall wilfully neglect to hold the meeting specified in the last preceding section, each assessor so neglecting shall be liable to a penalty of twenty dollars, to be sued for and recovered before any court having jurisdiction there f, by the supervisor of the town, for the use of the poor of the same town; and in case of such neglect to meet for review, any person aggrieved by the assessment ofthe assessors may appeal to the board of supervisors, at their next meeting, who shall have power to review and correct such assessment.

§ 6. Whenever any person on his behalf, or on behalf of those whom he may represent, shall apply to the assessors of any town or ward to reduce the value of his real and personal estate, as set down in their assessment roll, it shall be the duty of such assessors to examine such person under oath, touching the value of his or their said real or personal estate, and after such examination they shall fix the value thereof, at such amount as they may deem just, but if such person shall refuse to answer any question to the value of his real or personal estate, or the amount thereof, the said assessors shall not reduce the value of such real or personal estate. The examination so taken shall be written, and shall be subscribed by the person examined, and shall be filed in the office of the town clerk of the town or city in which such assessment shall be made, and any person who shall wilfully swear falsely on such examination before the assessors, shall be deemed guilty of wilful and corrupt perjury.

§ 7. The assessors of the several towns and wards of this State shall have power to administer oaths to any person applying to them under the provisions of the sixth section of this act.

§ 8. When the assessors, or a majority of them, shall have completed their roll, they shall severally appear before one of the justices of the town or city in which they shall reside, and shall severally make and subscribe before such justice, an oath, in the following form:

We, the undersigned, do severally depose and swear that we have set down, in the foregoing assessment roll, all the real estate situated in the (town or ward, as the case may be,) acording to our best information; and that, with the exception of those cases in which the value of the said real estate has been changed by reason of proof produced before us, we have estimated the value of the said real estate at the sums which a majority of the assessors have decided to be the full and true value thereof, and at which they would appraise the same in payment of a just debt due from a solvent debtor; and also that the said assessment roll contains a true statement of the aggregate amount of the taxable personal estate of each and every person named in such roll, over and above the amount cf debts due from such persons respectively, and excluding such stocks as are otherwise taxable, and such other property as is exempt by law from taxation, at the full and true value thereof, according to our best judgment and belief. Which oath shall be written on said roll, signed by the assessors, and certified by the justice, and shall be in place of the official certificate now required by law; and every assessor who shall wilfully swear false in taking and subscribing said oath, shall be deemed guilty of, and liable to the penalties, of wilful and corrupt perjury.

§ 9, This act shall take effect immediately.

AN ACT

In Relation to Railroad Corporations.

Passed February 13, 1851.

The People of the State of New York, represented in Senate and Assembly, do enact as follows:

§ 1. Whenever two railroad companies shall, for a portion of their respective lines, embrace the same location of line, they may, by agreement, provide for the construction of so much of said line, as is common to both of them by one of the companies, and for the manner and terms upon which the business thereon shall performed. Upon the making of such agreement, the company that is not to construct the part of the line which is common to both, may alter and amend its articles of association, so as to terminate at the point of intersection, and may reduce its capital to a sum not less than ten thousand dollars for each mile of the road proposed to be constructed in such amended articles of association.

§ 2. Whenever, after due examination, it shall be ascertained by the directors of any railroad company, organized under the act entitled an "act to authorize the formation of railroad corporations, and to regulate the same," passed March 26, 1848, or under the act entitled an "act to authorize the formation of railroad corporations, and to regulate the same," passed April 2, 1850, that a part of the line of their railroad proposed to be made between any two points in this state, ought to be located and constructed in an adjoining state, it may be so located and constructed, by a vote of two-thirds of all the directors, and the sections of the said railroad, within this state, shall be deemed a connected line according to the articles of association, and the directors may reduce the capital specified in their articles of association, to such amount as may be deemed proper, but not less than the amount required by law for the number of miles of railroad to be actually constructed in this state.

§ 3. Any railroad company formed under the act entitled, an "act to authorize the formation of railroad corporations," passed March 26, 1848, and which is duly continued in existence, when at least ten thousand dollars for every mile of its railroad proposed to be constructed in this state, shall be in good faith subscribed to its capital stock, and ten per cent. therof paid in may apply to the court for the appointment of commissioners, and the court shall thereupon appoint commissioners, and all subsequent proceedings may be had to obtain the title to lands necessary for its construction to the same extent and in the same manner, as if the whole amount of the capital stock specified in its articles of association was in like manner subscribed.

§ 4. In case any railroad shall occupy or cross any turnpike or plank road, the railroad company shall pay such turnpike or plank road company, all damages the turnpike or plank road company may sustain by reason of the occupancy or crossing such turnpike or plank road, the damages to be ascertained and paid in the same manner as is provided by law for the asssessment and payment of damages in case of taking private property for the use of railroad companies.

§ 5. This act shall take effect immediately.

NOTES.

The Laws in relation to Assessments, was passed April 15, 1851.

RAILROADS.

An action will lie against a R. R. Co., by a Religious Society, for a nuisance in running cars, blowing off steam &c. and thereby making noises during public worship, so as to annoy and molest the congregetion worshiping, and rendering the place unfit for public worship. *Baptist Ch., Schenectady, vs. S. & Troy R. R. Co.* 5 *Barbour*, 79. *Contra—See* 5 *Barb.* 70.

To sustain an action by an individual against a R. R. Co., for running over his wagon in crossing a railroad track, it must appear that the defendant's agents were guilty of negligence, and that the plaintiff was free from negligence, or fault. *Spencer vs. U. & S. R. R. Co.* 5 *Barb.* 387.

Appraisers appointed to value land taken by a railroad company, are not authorized to appraise the land of an individual with a reservations of easements and privileges of the owner; they must appraise the land at its actual value. *Hill vs. The M. & H. R. R. Co.*, 5 *Denio*, 206.

And where the inquisition stated that the award of damages was based on the supposition, and made on the condition, and with the understanding, that the owners of the might open a street, across the railroad—held on *certiorari*, brought by the owners that appraisement was illegal, and that the inquisition should be set aside. *Same.*

When Railroal corporation built the railroad across a turnpike without making or tendering any compensation therefor, it was held that though their charter declared it lawful for them, when necessary, to lay their road across any highway, they were liable to the turnpike company, for damages. *Seneca Road Co. vs. A. & R. R. C.*, 5 *Hill*, 170.

As to the rights of a company to build their track in the streets of a city, by leave of the corporation of the city. *See Drake* vs. *Hudson R. R. Co.*, 7 *Barb.*, 694.

In an action against a R, R. Co., if appear that through the negligence of the persons runniug the cars, an accident was about to take place, and the plaintiff had to jump out of the cars to save his life, and was injured; it was held that he could recover. *Eldridge* vs. *the Long I. R. R· Co.*, 1 *Saun. S. C.* 89.

The privilege of making a R. R., and taking toll thereon, when granted to an individual or a company, is a *franchise.* The public have an interest in the use of the road, and the owners of the franchise are liable to respond in damages, if they refuse to transport an individual or his property upon such road, without any reasonable excuse, upon being paid the usual rate of fare. *Beekman vs. Saratoga and S. R. R. Co.*, 3 *Paige*, 45.

An action lies by the owner of land against a R. R. Co., if in the construction of their road upon *or across a pulic highway*, they raise embankments, by which the owner of the land is obstructed in passing to and from the road, and his property is otherwise rendered less valuable, notwithstanding the charter of the company authorizes the entry upon, and use of, such public highway; the license relates only to the road, and leaves the company liable to consequential damages, sustained by individuals. *Fletcher vs. Auburn & S. R. R, Co.* 25 *Wend.* 462.

ASSESSMENTS AND TAXES.

A person entitled to receive *rents* subject to taxation by the act to equalize taxation passed May 12, 1846, has a right to avail himself of his own affidavit as to the value of his property, under the provisions of the Revised Statutes for the purpose of correcting an assessment upon such *rent.* *Livingston vs. Hollenbeck*, 4 *Barb. S. C.* 9.

The general law is unrepealed, and that and the act of 1846, must be construed together, and it is the policy of the law that all valuation assessments may be corrected by the affidavit of the person assessed. *Same.*

When such an affidavit is presented correct in form, it is the duty of the assessors to correct their assessments according to the § 15 of R. S., as to assessments, and they have no discretion to reject the affidavit when tendered by the person assessed. They are bound to receive and strike out the assessments accordingly.—*Same.*

The assessors in estimating the value of property, act *judicially* and if their judgment is honestly exercised, it is conclusive upon all interested, except so far as the affidavit of the owner of the premises may serve to correct it.—*Same.*

The assessment of property is a judicial act upon which a certiorari will lie. But to make the assessment legal, the assessors must have jurisdiction of the particular case. If they transcend the limits of their authority and undertake to assess property exempt by Statute, they cease to be judges and are responsible for all the consequences. *Prosser, vs. Secor, Barb.* 607.

They have no right to enter upon the assessment roll the name of a person exempt by law ; and accordingly *held* that they had no right to enter the name of a Minister of the Gospel not having property over $1500, and in determining whether they have jurisdiction or not they do not act judicially.—*Same.*

In order to take a man's unoccupied land from him, for the non-payment of a tax, the law requires, that if the township patent or tract of which it is a part has been subdivided into lots, such land shall be assessed, by the *number* of the lot, if it can be ascertained. *Dike, vs. Lewis,* 2 *Barb.* 344.

A suit cannot be maintained against the assessors of a town, for refusing to give the plaintiff, who is a taxable inhabitant, the benefit of an exemption, nor for assessing property at too high a rate when they have jurisdiction.— *Weaver, vs. Dievendorf,* 3 *Denio* 117.

A town collector may seize not only the goods and chattels of the person taxed, but any goods and chattels in his possession.—*Sheldon vs. Van Buskirk,* 2, *Com.* 473.

When a collector levies on goods before the return day specified in his warrant, he may, under the Statute (2 R. S. 398, § 6,) *sell* at any time within a week after such return day.—*Same.*

A tax is fatally defective that describes the lot in a different tract from that in which it is situate. A Comptrollers deed under it, conveys no title.—*Tallman vs. White,* 2, *Com.* 66.

INDEX.

A

B

C

D

www.ingramcontent.com/pod-product-compliance
Lightning Source LLC
LaVergne TN
LVHW050517100826
845148LV00002B/362

* 9 7 8 1 4 2 5 5 2 1 4 2 4 *